CW01511686

GRAPE EXPLICATIONS

"I have enjoyed a lot of what I have read, and particularly the titles of some of the fabulously iconoclastic papers delivered to the conferences. [Hulkower has] a very lively mind and that is always on display, right the way through to the swashbuckling 'Devil's Wine Glossary' at the end. The book is dauntingly detailed yet thoroughly entertaining—a rare combination in the wine world."

—ANDREW JEFFORD, author of *Drinking with the Valkyries*

"As in his search for elegance both in mathematics and wine, Neal's writing keeps us present and in the moment. In his endeavors for the *Slow Wine Guide USA*, he has expanded our collective awareness and appreciation of the wines of Oregon and their makers."

—DEBORAH PARKER WONG, National Editor of the *Slow Wine Guide USA* and Global Wine Editor of *THE SOMM JOURNAL*

"This is a delightful book. Neal writes with rigour and balance, and his style is easy to read."

—JAMIE GOODE, author of *I Taste Red* and *The Science of Wine*, and Chief Anorak, wineanorak.com

"Neal's long and varied path in Oregon wine leaves him uniquely suited to bring insight to this unique time and place in our community."

—JASON LETT, proprietor and winemaker, The Eyrie Vineyards

"In *Grape Explications,* Neal Hulkower takes us by the hand and leads us through his own diverse, eclectic, and intelligent journey through the world of wine in this highly personal collection of insights. They are every bit as wide-ranging, enlightening, and fun as if you were conversing, glass in hand, across a tasting table with the author himself. Hulkower rambles from a 'Stellar Bash in Stellenbosch' to 'risking life and limb' in far-off Georgia. He has collated data on vintages, dives into the somewhat arcane world of wine economists, and offers candid reviews of some of the seminal wine books of recent years. Along the way, Hulkower is unafraid to 'call a spade a spade' when 'underwhelmed' by a wine offered for review. He is most at home when dealing with the locals in his hometown in Oregon's Willamette Valley. His keen and unabashed appreciation of pinot noir shines through. Written in a clear, concise style, there's a little something here for everyone who shares the author's passion for the myriad, multifaceted beauty contained in a glass of wine."

—**DOUG TUNNELL**, founder and vintner, Brick House Wines

"A mordant romp through a well-lubricated life by mathematician-turned-wine writer Neal Hulkower. Cogent, sometimes contentious, and unfailingly witty, Hulkower's elegantly organized tome illustrates that the sum of parts of a lifetime contemplating fine wine indeed proves greater than the whole."

—**L.M. ARCHER**, award-winning wordsmith

NEAL D. HULKOWER
Foreword by Jamie Goode

NDH PRESS

Grape Explications
Copyright © 2025 by Neal D. Hulkower

All rights reserved. This book or any portion thereof may not be reproduced or used in any manner whatsoever without the express written permission of the author, except for the use of brief quotations in a book review.

Printed in the United States of America
Cover design by Glint Creative
Interior icon by Freepik

NDH Press
McMinnville,Oregon

LCCN: 2024922336
ISBN: 979-8-9903142-0-7

Dedicated to Don Saari, who first suggested I compile some of my wine pieces into a book and persisted until I did, and to my ever-supportive wife, Clara, who did the first editing of most of what is included here

TABLE OF CONTENTS

Foreword

I first came across Neal Hulkower through his book reviews. As a book author with a reasonable steady output, one of the things that delights me is when someone takes the trouble to do a thorough but fair book review. I've been on the end of Neal's thoughtful, insightful, and generous reviews a few times, and a couple of these reviews are included here in the section on wine books.

Neal's book is broken into different parts, gathering together his writing in sensibly arranged topical themes. As well as the aforementioned book reviews, there's material based on Neal's involvement with the American Association of Wine Economists, plenty on the Willamette Valley in Oregon where he now lives, and a section on wine personalities. He's clearly a wine geek, but while he's writing about wine, he keeps the necessary distance between himself and the wine trade, and that helps give some objectivity to the work. He's writing for the reader, not the people or wineries that he's featuring.

This is a delightful book. Neal writes with rigor and balance, and his style is easy to read. Compilations like this are great: articles worthy of attention would be dispersed and forgotten without being brought together in this way between two covers. I wish more writers would do this with their work. Yes, some of the reports are no longer current, but they give a snapshot of what was happening when they were being written. Reading this book is like popping up to an elderly relative's attic to open old chests full of family artifacts. There are treasures here waiting to be found.

—Jamie Goode

Preface

Finally, the time has come. After embarking on a sideline gig as a freelance wine writer in 2011, I have accumulated enough pieces to more than fill a book. I was encouraged to do so by Don Saari, who had been my dissertation advisor at Northwestern and later became a fan of my more recent work.

This compilation has no single theme. Several have emerged, however, and serve as the titles of the parts. Since this is the work of a single author, elements one might expect in a memoir are present. My background as an applied mathematician also surfaces in several of the pieces. My initiation into the world of fine wine and the various vinous adventures I had over the years made for popular pieces that are included.

I started with the urtext and lightly edited the entries. I also added annotations and updates when necessary since in more ways than one, wine is a fluid subject. Several entries have not been published or posted anywhere before. The articles are organized into eight parts.

"In the Beginning" introduces my earliest wine writings and also includes an essay about my first interactions with German wines. Around the same time that I left full-time employment and immersed more fully into the wine world, I joined the American Association of Wine Economists (AAWE) since there is no equivalent for mathematicians. The annual meetings are held in wine regions around the world, giving me an excuse to travel to places I wouldn't necessarily have visited, as well as providing a venue for sharing some of my research. Part II comprises my reports on these meetings.

As part of my deep immersion in wine, I expanded my reading on the subject. In addition to mostly online publications and other resources, I sought out books. In order to feed my habit and reduce costs, I began writing book reviews for two academic publications, the *Journal of Wine Research* and AAWE's *Journal of Wine Economics*. If you are reading this (thank you!), you might search for some of the titles included in the third part.

I became a full-time resident of McMinnville, Oregon, the heart of the Willamette Valley, in July 2011. Three months later, my first piece appeared in the *Oregon Wine Press* (OWP). It has since proven to be receptive to many of my pitches. I include a few along with others that haven't appeared elsewhere and one that did in Part IV. I hope that they give a flavor of what life is like in this wonderful and burgeoning wine region. I have also met numerous wine personalities. Part V includes articles about four of them.

The *Asociación Hispana de la Industria del Vino en Oregon y Comunidad* (AHIVOY), the brainchild of Jesús Guillén, Sofía Torres McKay, and Miguel Lopez, holds a special place in my heart. At Jesús's request, I first wrote about it when it was still an inchoate idea to pay forward to the vineyard stewards the good fortune the three had experienced. I continued to chronicle the realization of the idea and the early successes. When I was invited to join the board in 2022, I handed over the task of publicizing the happenings to another writer. Part VI chronicles the early years of the organization.

In 2019, I was recruited to become a field coordinator for the *Slow Wine Guide USA*, an offshoot of the Italian edition sponsored by the Slow Food movement. While the remuneration is meager—so much so that it almost qualifies as pro bono work—I continued to participate. I would be lying if I said that I didn't take pleasure in going around and tasting remarkable juice. Some of the winemakers I visit each year have come to expect and enjoy our sessions, and the ones who I add appreciate the exposure. To add a bit of revenue and to publicize the guide, I published two reports in the OWP about the wineries I covered which constitute Part VII.

The book title is taken from an essay I published in the OWP on the fifth anniversary of the appearance of my first article in that monthly. I have since been successful in placing a range of pieces in other outlets as well. Part VIII, which also carries the name of the book, highlights some that I particularly like or that have especially pleased my readers.

Several of the books I've reviewed contain a glossary at the end. Instead of producing yet another traditional one, I decided to take a cheekier approach. Let's just say that the devil made me do it.

Feel free to jump around and sample the essays in whatever order you fancy. If they amuse, entertain, educate, encourage you to try a new bottle, or open a new book, I will have succeeded.

L'Chaim
Neal Hulkower
McMinnville, Oregon
February 2024

PART I

In the Beginning

The Notebooks

Between 1969 and 1979, the first decade of my growing obsession with wine, I kept notes of almost everything I drank. These are contained in four small loose-leaf notebooks. Each page includes the name and vintage of the wine, its producer and importer, the price paid, the date it was consumed, and extensive tasting notes. There is an "L" in the upper-right corner of the page if I had kept the label. All told, 450 notes were compiled, some reflecting multiple tastings of the same wine at different times.

The number of wines from each country including multiple tastings of the same wine are 254 from France, 124 from Germany, 38 from the United States, 21 from Portugal, 6 from Hungary, 4 from Spain, 2 from Italy, and 1 from Argentina. These became the source material for articles and studies, some of which are included in this volume. The notebooks are also a source of amusement and amazement for guests to whom I have shown them.

Here are the first of my wine-related publications. They appeared in the now-defunct Vintage *magazine, which solicited tasting notes from readers. I was happy to oblige.*

When I was an undergraduate and had just discovered that food and wine were supposed to taste good, I cofounded a gourmet group that we named the Duncan Hines Memorial Bon Vivant Fellowship, Int'l. It thrived during my graduate school years sponsoring an annual banquet and tastings of mostly French and German wines.

The first set appeared in the September 1973 issue. The tasting notes are from my master's degree dinner on June 17, 1973. Our tradition was to have an Auslese when getting a bachelor's degree, a Beerenauslese for a master's, and a Trockenbeerenauslese for a PhD.

The *Vintage* Tasting Notes

From Neal D. Hulkower
Minister of Wine
Duncan Hines Memorial Bon Vivant Fellowship, Int'l.
Evanston, Illinois:

Chateau Carbonnieux Blanc 1970: Said to be the finest white wine ever produced by this chateau, it was certainly the best vintage I had tasted (the others being 1964, 1967, and 1969). Characterized by an abundance of fruit in the bouquet, this Graves had the fullest body and best balance ever noted for its kind. The silvery straw color was deeper than expected. The texture was fine and the aftertaste was clean, long-lived, refreshing, and elegant. This wine lent its most inviting characteristics to the Sauce Parisienne to which it was added. Available in Chicago for $4.39.

Chambertin 1947, Ponnelle: An astounding wine with a smell (bouquet is much too mild a word) of truffles. Further descriptions of the nose included austere, powerful, and heady; mushrooms and beef were detected at opening. Rivaling the perfume for attention was the color, which is best described as a deep purple robe with no signs of age. Clearly, it was the most opulently deep-red color seen by anyone in a burgundy. The texture was close-knit and velvety, although less than expected. The taste and aftertaste were round and full but didn't evolve much and weren't related to the bouquet. A hint of mint and flowers was detected by one taster, but the fruit was clearly replaced by truffles. This wine had more depth and less fruit than a 1964 Remy Chambertin previously tasted but gloriously fulfilled the general description of a Chambertin in Lichine's *Wines of France*. Available in Chicago for $19.

Hattenheimer Heiligenberg Feine Beerenauslese 1969, Nitzling: This little-known lager produces surprisingly fine fruity wines. While thinner and less rich in body for either a 1969 or a Beerenauslese, this Rheingau was a veritable liquid fruit bowl with peaches, apricots, pears, and persimmons detected in the bouquet. While the color was a lighter-than-expected yellow, the balance between fruitiness, sweetness, and acidity was beautiful. The texture was smooth but not thick, like a luscious nectar. Less *würzig* than expected, but quite excellent for what it was. Available in Chicago for $25.

Grande Fine Champagne Cognac, V.E., M. Ragnaud: This outstanding cognac deserves special consideration. It is the least aged and hence least expensive cognac produced and estate-bottled by M.

Ragnaud, but it beats hands down any commercial brand in its class. The taste is initially austere, with no bite, and quite dry, but develops and rounds into a lovely grape flavor. The bouquet is balanced and grapey. The color is medium brown with ever so slight golden over-tones. Available in Chicago for $8.99.

The second set was published in the January 1974 issue. The description of the event is given prefacially to the tasting notes. I'm almost certain that we used the Borda count to arrive at the consensus ranking well before we knew what it was called and years before my dissertation advisor Donald Saari proved theorems about it long after I finished my degree.

From Neal D. Hulkower
Minister of Wine
Duncan Hines Memorial Bon Vivant Fellowship, Int'l.
Evanston, Illinois:

Enclosed are notes recorded by eight members and guests of the fellowship who compared 1971 vintage Spätlesen from five of the best vineyards of the Rheingau on September 29, 1973. All prices are those paid in Chicago. Mr. Peter G. Rudiger of Schaeffer's, Skokie, Illinois, assisted in planning the tasting and supplied some of the wines. Color and general body will not be emphasized in the notes for the individual wines since these characteristics were virtually duplicated in the five bottles.

The color was pale yellow, described as straw or gold depending on the individual taster, the background on which the wine was viewed and the individual wine, with occasional green overtones, noted. The wines differed noticeably but not significantly in the shade of yellow.

It must be emphasized that each of these wines manifested a full-bodiedness that elevated them orders of magnitude above lesser vintages, such as 1969, and established them in the realm of the greats. That is not to say there was no range of body and texture; these will be compared in the notes below. However, each wine displayed its distinct essence, fully and unequivocally.

Finally, although we have ranked these five wines, these rankings were purely relative and not intended to deprecate any vineyard. Not a single bottle was off; not one wine was a disappointment.

Erbacher Marcobrunner Spätlese 1971 (State Domain), $5.98: Abundantly fruity, balanced bouquet, manifesting a complex blending of peaches and pears. Very sweet, much more than the average Spätlese and closer to an Auslese, and fruity taste with superb balance, acid providing the counterpoint. Very clean and easy to drink. No characteristic spiciness. Silky soft, velvety texture. Taste finished honeyed. The wine is still a little immature but very pleasant for present consumption. Ranked second.

Steinberger Spätlese 1971, $5.49: A powerful, fruity-spicy (cinnamon) nose, but not as fruity as the Marcobrunner. An intense, powerful taste, complex and spicy, develops into a heroic tart aftertaste. The Steinberger seemed to better characterize the quintessence of the Rheingau. Ranked first.

Rauenthaler Baiken Spätlese 1971 (State Domain), $5.98: Strong, fruity, spicy bouquet, very complex. Some thought the balance to be tipped in favor of spice; others toward fruit. Deep, potent, spicy, honeyed taste noted by one person. Tart and slightly sour in aftertaste, which starts fruity and somewhat sweet. Ranked fifth.

Schloss Vollrads Spätlese 1971, $6.79: Small nose but beautifully balanced bouquet, clean and fresh. Balance in taste struck between acid and fruit. Noble. Spice develops during taste and aftertaste. Medium to full body. Perhaps not as developed as others. Ranked fourth.

Schloss Johannisberger Grunlack Spätlese 1971, $5.49: Rich, elegant, forthright bouquet. Silvery polished texture, lightest in body of the five but perhaps the most perfectly balanced. Deep, fruity, slightly sweet flavor with pleasant long-lived distinguished aftertaste. *Hochedel.* Ranked third.

Here is the first wine-related piece I published since the late 1970s. It is a "Golden Book" version of my article that appeared in the Journal of Wine Research *in 2009 that describes the Borda count, a method of aggregating rankings that uniquely satisfies four rational properties thereby best reflecting the preferences of the voters. Since many wine competitions yield a ranking of the wines sampled, the Borda count is well suited for arriving at the consensus. While it wasn't used for the original Judgment of Paris tasting in 1976, it was employed thirty years later when the competition was held again with the same wines.*

Borda is Better

In 1959, Professor Maynard Amerine and his colleagues developed the "University of California at Davis 20 Point Scale System Organoleptic Evaluation Scoring Guide for Wine" or Davis Scoring System, for short. Though it was originally intended to rate experimental wines being made at that venerable institution, it or some variant has been widely adopted as the basis for ranking wines in competitions. The famous Judgment of Paris tasting in May 1976, which placed a California cabernet sauvignon and a California chardonnay in first place over several of France's best bordeaux, was decided by adding the scores assigned by nine judges on a 20-point scale. The *Oregon Wine Press* tasting panel also uses a 20-point scale in determining a rank-ordered list of recommended bottlings. But is this the most defensible way?

Actually, allowing individuals to assign scores on a 20 or 100 or any other point scale and determining the best by adding them is fundamentally flawed and can result in outcomes that do not best reflect the preferences of the tasters. As Orley Ashenfelter and Richard Quandt noted in a 1999 article in *Chance*:

"The problem with this approach is, of course, that it may give greater weight to judges who put a great deal of scatter into their numerical scores and thus express strong preferences by numerical differences."

Simply put, easy graders' votes would carry more weight in the tally than tougher ones, thereby violating the sacred principle of "one taster, one vote."

One way to preserve the sacred principle is to have each taster simply rank-order the wines and then use what is called a positional voting method to aggregate the rankings, yielding what is called the societal outcome.

Let's look at an example. Suppose thirteen tasters gather to select one of three wines to be served with the main course at a dinner. The choices are Amazing Abbey (A), Bodacious Bodega (B), and Chic Château (C). The table below gives the number of tasters voting for each of the six possible rankings of the three wines:

Number of Tasters	Ranking of A	Ranking of B	Ranking of C
4	1st	2nd	3rd
2	3rd	1st	2nd
1	2nd	3rd	1st
2	1st	3rd	2nd
0	2nd	1st	3rd
4	3rd	2nd	1st

Now what? How can we aggregate this profile to arrive at the societal outcome that best reflects the preferences of the tasters? Here are three of the infinite ways to do so.

First, we can implement plurality voting by assigning one point to the first-place wine and zero points to the others. Amazing Abbey is the plurality winner with six points, with Chic Château coming in second place with five points, and Bodacious Bodega in third with two points.

Next, we can try the antiplurality method by assigning one point to all but the last placed wine, which gets zero points. This is voting *against* one option as opposed to *for* one using plurality voting. The antiplurality winner is Bodacious Bodega with ten points followed by Chic Château with nine points, and Amazing Abbey with seven points.

Finally, we can look at a method called the Borda count that is, in some sense, in between the extremes of plurality and antiplurality voting. Borda assigns a score of two to the first-place wine, one to the second, and zero to the third. The Borda winner is Chic Château with a Borda score of fourteen, followed closely by Amazing Abbey with thirteen, and Bodacious Bodega with twelve.

Hmmm. What we see is that societal outcome depends on the method used to aggregate the individual preferences. Have we just made matters worse? Do we have a way of knowing which of the infinite number of methods of combining the rankings—or even which of the three above—best reflects the preferences of the tasters?

Fortunately, mathematics has come to the rescue. Prof. Donald Saari of the University of California at Irvine has proven that the Borda count is unique among all possible positional voting methods in that it satisfies four simple and rational criteria.

In addition, it is less likely than any other positional voting scheme to experience paradoxes or outcomes that are inconsistent with the voters' stated preferences. Lest I induce a mathematical hangover, I'll refer those interested in the details to Saari's 2008 book, *Disposing Dictators, Demystifying Voting Paradoxes, Social Choice Analysis*, Cambridge University Press, which contains a detailed but reader-friendly explanation of the virtues of Borda.

Implementing the Borda count is easy. If there are n wines being tasted, the first-ranked bottle receives a score of n-1, the second, n-2, and so on, with the wine ranked last getting a score of 0. In the case of ties, each wine receives the average of the scores assigned to the rankings the group occupies. So, for example, if three wines out of ten are tied and occupy the third through fifth positions, each would get a score of $(7+6+5)/3 = 6$. The scores for each wine are summed to determine the societal outcome.

In a paper published in 2009 in the *Journal of Wine Research*, I reexamined the Judgment of Paris results. When I converted the points assigned by the nine French judges to rankings and used the Borda count to aggregate the results, Château Haut-Brion 1970 emerged in first place whereas the declared winner, Stag's Leap Wine Cellars Cabernet Sauvignon 1973, dropped to second.

The declared winner of the chardonnay competition, Chateau Montelena Chardonnay 1973, a California wine, was also the Borda winner. In 2006, thirty years after the original competition, the red wine tasting was done again. This time, the Borda count was used to aggregate the results of two panels. In first place was the Ridge Monte Bello Cabernet Sauvignon 1971, with Stag's Leap second and Haut-Brion eighth. Complete results can be found at http://www.vinography.com/archives/2006/05/the_rejudgment_of_paris_result.html.

I have been spreading the gospel according to Borda and have found some converts. In July 2011, Harry Peterson-Nedry asked Chehelam Tasting Panel members to rank-order three candidates for the Tasting Panel Cuvée 2010 and used Borda to determine the societal outcome. There was a tie for first place between a fuller, richer blend and one that had all of the same elements of the other but was more accessible. Harry broke the tie by selecting the bigger wine since it had more first-place votes.

Later that same month, I was one of six tasters who ranked five vintages of Archery Summit Renegade Ridge Estate Pinot Noir. The societal outcome according to Borda in descending order was: 2001, 2002, 2004, 2003, and 2000. All were excellent and would benefit from additional aging.

This is one of the articles I wrote using data mined from The Notebooks. It celebrates a period when I drank German rieslings with much greater frequency than I do now.

A Remembrance of German Wines Past

My twenty-something self left two gifts for the older guy I would become—a doctorate in applied mathematics and four notebooks. The degree was a credential that not only opened many doors but also, in its attainment, reinforced my ability to do independent research, perform analyses, and document the results lucidly. The four notebooks, along with labels from many of the bottles, form an archive of my first decade of tasting fine wine.

My fascination with wine developed in parallel with my maturation as a professional mathematician. Decades later, these two interests intertwined with a foreshadowing of how this would happen while I was still a student.

From 1969 to 1979—the period covering the end of my undergraduate years, my entire time in graduate school, and my first job after I graduated—I kept detailed notes on almost everything I tasted. With a group of like-minded students, I cofounded the Duncan Hines Memorial Bon Vivant Fellowship, Int'l, one purpose of which was to hold tastings. These provided regular opportunities to sample more and better wines more frequently.

The wines that hooked me over half a century ago were mostly French and German. These were both more readily available in the Chicago area market and recognized as among the highest quality. Of the 450 records kept in those small loose-leaf notebooks, 254 are of French wines and 124 are German, some of which were tasted on more than one occasion. In an era when Liebfraumilch, Zeller Schwarze Katz, and Blue Nun would have been more affordable on a student's budget, I was drinking Prädikatsweins from such places as Wehlen, Hattenheim, Wiltingen, and Ockfen.

Wine, in general, wasn't common within my college community. There was a vibrant food scene in Chicago, with its broad collection of ethnic and fine dining restaurants, which certainly encouraged wine consumption along with the meal. On the other hand, I was in Evanston, which was dry, but there were several stores in the area that had excellent selections of fine international wines.

The earliest entry of a German wine in the notebooks is the 1966 Wehlener Sonnenuhr Auslese from S. A. Prüm, tasted in May 1970. The price was around $7. My terse description reads "lightly sweet, fruity, delicate" and notes that it is a selection of the DHMBVF, Int'l wine committee. This vineyard provided fifteen of the wines that are in the notebooks. In addition to S. A. Prüm, C. Prüm Erben, Bergweiler-Prüm, Heinrich Ludwig Bäumler Erben, Erich Friedrich-Adams, and a particular favorite, Joh. Jos. Prüm, appear as producers. The latter's wines were often more complex, especially floral, and more consistent at balancing sweetness, fruitiness, and acidity.

On June 17, 1970, shortly after getting my BA, I drank the 1966 Winkeler Hasensprung Auslese from A. von Brentano at the now-shuttered and demolished Lüchow's Restaurant in New York. I have no recollection of the meal, but the wine was "beautifully balanced, [with a] very delicate light bouquet [and] fine color, called 'Goethewein.'" The price was $9.

My master's degree dinner was exactly three years to the day later. With the help of a friend, I prepared a six-course meal. The dessert, a Souffle Rothschild, was paired with the 1969 Hattenheimer Heiligenberg Feine Beerenauslese from Weingut Nitzling ($25, about the same as the cost of a week's worth of groceries for two): "Light in body and color, but beautifully balanced in fruit and sweetness. Not a luscious nectar, but a pleasantly asserting light liqueur. Clear medium-yellow."

The DHMBVF organized a tasting of six 1971 Auslesen from the much-heralded 1971 vintage on June 29, 1974. Admittedly, they were young but so were we, and impatient as well. Our practice was to rank the wines individually, and then aggregate the rankings to arrive at a group consensus. My recollection is that we did so by simply adding the individual rankings and awarding first place to the wine with the smallest number and proceeding down to the next lowest for second place, and so on.

Little did I know at the time that how best to aggregate rankings was a rich mathematical problem that my dissertation advisor, Don Saari, would solve over a decade after I finished graduate school. It turns out that the Borda count, which is equivalent to what we did, uniquely satisfies four rational properties. In recent years, I have applied the Borda count to numerous tastings and published results.

The results of the tasting are below:

My Rank	Group Rank	Wine	Producer	My Notes
1	2	Wehlener Sonnenuhr Auslese	S. A. Prüm	Light straw, some green overtones. Delicate perfumed fragrance. Good acid, full taste flowery after bouquet. Good aftertaste, long-lived.
2	1	Bernkasteler Graben Auslese	Pfarrkirche	Light straw color, green overtones. Full noble taste, good acid. Under-developed nose, but nice. Buttery aftertaste (finish) and texture.
3	4	Eitelsbacher Marienholz Auslese	Bischöflicher Konvikt	Light straw color. Full, bold nose with fruit and some flower apparent. Medium taste after the nose. Steely finish. Long-lived, nonevolving aftertaste.
4	5	Ockfener Bockstein Auslese	Verwaltung der Staatlichen Weinbaudomänen	A delicate floweriness appears in the nose with some hint of apple or fruit with taste after bouquet, light straw color. Some steel, but not as much as expected.

5	3	Piesporter Gold-tröpfchen Auslese	Bischöflicher Konvikt	Light straw, some green overtones. [Sulfurous] bouquet, at first, some flowers, seems underdeveloped. Good acid in taste. Some steeliness (about the same as Eitelsbacher Marienholz).
6	6	Graacher Himmel-reich Auslese	Kunibert Flesch	A strange, less than inviting bouquet of St. John's Bread. Flat, short-lived taste after bouquet. Doesn't last, hits and dies. Can't see a future for it.

On November 2, 1974, we drank a 1949 Wehlener Sonnenuhr Feinste Auslese from J. J. Prüm as part of an early twenty-fifth birthday celebration of that same friend who helped me with the master's degree dinner and also shared the same birth year with me. It was purchased for $25 in 1972. The notes are: "Fine gold color. Rhinegau (sic) nose [yes, I know it is from the Mosel]—spicy, honeyed fruit, forthright. 'Fat' taste, lacking acid, but flavor is after the nose, with long, noble aftertaste, somewhat sweet and spicy. Glorious."

On May 22, 1977, a tasting of classified growth clarets was capped with the 1959 Steinberger Trockenbeerenauslese to celebrate completing and successfully defending my doctoral dissertation. I was told by George Schaefer, proprietor of his eponymous wine shop in Skokie, Illinois, where a significant portion of my stipend went, that the wine had just gone for $200 at auction. However, he was willing to part with his only bottle for $85 if I would pour off an ounce for him. It was worth every penny:

"Deep dark brownish apricot—gold. Rich intense fruity aroma—thick. Remarkable flavor—rich, thick, oily, cushy texture. Initially, a very pleasant acid tinge—could be the bubbles—followed by the honeyed fruit and a depth of finish never before experienced—maintains and glorifies the family traits exhibited by the 1971 [Steinberger] Spätlese [which began the tasting]. Actually showed some secondary fermentation. Nectar with good balance, sufficient acidity—fantastic."

I recall that those in attendance grew silent when they first sipped this wine. To me, it was like biting into a ripe apricot. Years later, I could still imagine this sensation. Sound good? You'll now (in February 2021) need $2,848 to try it yourself. This is a relative but still unattainable bargain compared to $3,610, the price in May 2020.

Not everything we tasted was made from riesling. In 1975, I had a 1973 Westhofener Steingrube QbA, a blend of silvaner and müller-thurgau, and a 1964 Westhofener Kirchspiel Rulander Riesling Beerenauslese, both from J. G. Orb.

The following year, I drank a 1972 Westhofener Moorstein Siegerrebe Beerenauslese, a 1974 Westhofener Steingrube Ortega Optima Auslese, both also from J. G. Orb, a 1971 Siebeldinger Königsgarten Morio Muskat Auslese from G. Pfaffmann, and a 1971 Guntersblumer Himmeltal Scheurebe Auslese from Schätzel Erben.

A 1971 Westhofener Moorstein Auslese from J. G. Orb is the most recently tasted German wine in the notebooks. The price was $6.98. On March 10, 1979, the third time I had it, I wrote: "Last bottle. Warm dark apricot. Intensely fruity nose—papaya, mangos. Mature deep rich. Flavor less intense, with a hint of acid."

In the summer of 1978, I was invited to present the results of my dissertation at a conference at the Mathematisches Forschungsinstitut Oberwolfach in the Black Forest. It was my first trip to Germany. Though I was no longer in the area and hence no longer a customer, George Schaefer kindly agreed to arrange a letter of introduction from Kendermann. This led to six days of touring and tasting in the Rheingau and Mosel. Unfortunately, my notebooks did not join me on the trip, so I must rely on faded memories.

Our first stop was Kendermann's office in Bingen where we enjoyed a hastily arranged tasting of about a half dozen wines. This was followed the next day with tours of the Schloss Johannisberg and Schloss Vollrads. We then sailed down the Rhine and made our way to Zeltingen for a weekend wine festival along *die schöne Mosel*. While it was certainly pleasant strolling the three streets of the typically tiny wine town, a small glass in hand for sampling whatever was being dispensed, most memorable was what came next.

The town of Wehlen is across the river and a short walk from Zelt-ingen. On Monday morning, we knocked on the door of the house on the label of J. J. Prüm wines since we thought that an appointment had been made for us by Kendermann. It had not.

Nevertheless, Dr. Manfred Prüm invited us in for a very brief visit since he had another appointment elsewhere. We were seated in a lovely salon and handed a glass. I took a sniff, inhaling the intense fruit, and without having the good sense to taste first, I blurted out, "Auslese!" to which Prüm immediately replied "No!"

Had I actually put some in my mouth instead of hotdogging it, it would have been obvious, and I likely could have identified the 1973 Wehlener Sonnenuhr Kabinett. The vintage was notable for richer Kabinetts but not for higher-level wines. Though my blind-tasting ability remains poor, I always taste first before committing to what is frequently the wrong answer.

On the way to his appointment, Prüm drove us to ours at the Thanisch cellars where I sampled the 1975 and 1976 Bernkasteler Doktor Auslesen. Although the wines were certainly excellent, at about $35 each, they were out of my price range.

While my younger self knew the benefits that could accrue from a PhD, they would be less obvious for the notebooks. Career and family pressures brought about the end of this record of wonderful wines past; however, delving into it from time to time has been a joy.

As my career advanced, I moved around a lot. When I lived in or near a wine region, I tended to focus on the local wines. Nevertheless, my love of riesling, first developed by drinking German wines, has only increased. While at this writing, there are no wines from Germany of any variety in my collection, I have rieslings from Oregon, the Finger Lakes of New York, and Michigan. Still, from time to time, I'll get a bottle of German riesling, and upon drinking it, wonder why I don't do it more often.

A 2018 Niederhauser Hermannshohle Riesling Grosses Gewachs from Dönnhoff is now part of my collection. I'm in no rush to try it.

PART II

American Association of Wine Economists

The American Association of Wine Economists and Me

Founded in 2006 along with the *Journal of Wine Economics* that it publishes, the American Association of Wine Economists (AAWE) offers its international membership the opportunity to share research on drink, food, and cannabis-related topics. There has been an annual meeting since 2007, interrupted only by the pandemic, convened near winegrowing regions around the world. These include sessions at which research is presented plus wine tastings and winery tours. The AAWE also invites researchers to submit working papers, less formal than a peer-reviewed journal article, describing work in progress or interim results. These are cataloged and made available to anyone on the website https://wine-economics. org/working-papers/

I joined the AAWE in 2012, submitted my first working paper, and attended my first annual conference in June of that year at Princeton University. Since then, I have produced three more working papers and gone to all but two annual conferences due to conflicts. At all but one, I gave at least one presentation, four of which were based on my working papers:

"A Mathematician Meddles with Medals," AAWE Working Paper No. 97, February 2012 and presented at the sixth Annual Conference at Princeton University in New Jersey in June 2012.

"Three Vignettes about Wine Tastings and Competitions," presented at the seventh Annual Conference in Stellenbosch, South Africa, in June 2013.

"Information Lost: The Unbearable Lightness of Vintage Charts," presented at the eighth Annual Conference at Whitman College, Walla Walla, Washington, in June 2014.

"Minimum Percent Error-Zero Percent Bias Regression for Wine Economists," presented at the ninth Annual Conference at the Universidad Nacional de Cuyo in Mendoza, Argentina, in May 2015.

With Stokes, S. Lynne, "Toward Valuing Willamette Valley Pinot Noir as a Cultural Good," AAWE Working Paper No. 245, January 2020 doi:10.13140/RG.2.2.27346.09924 and presented at the eleventh Annual Conference at the Botanical Garden in Padua, Italy, in June 2017.

"How to Decide How to Decide," presented at the twelfth Annual Conference at Cornell University in Ithica, New York, in June 2018.

"What Can I Still Afford to Drink?" AAWE Working Paper No. 254, June 2020 doi:10.13140/RG.2.2.25178.57281 and presented at the fourteenth Annual Conference at Tbilisi State University in Tbilisi, Republic of Georgia, in August 2022 (One of three awarded the Christophe Baron Prize for the Best Conference Presentation).

"Contre-degustation Olympiades du Vin According to Borda," AAWE Working Paper No. 254, September 2021 doi: 10.13140/RG.2.2.29320.75521 and presented at the fourteenth Annual Conference at Tbilisi State University in Tbilisi, Republic of Georgia, in August 2022.

"A Quality Price Ratio Comparison of Willamette Valley Chardonnay and White Burgundy," presented at the sixteenth Annual Conference at EHL Hospitality Business & Hotel Management School in Lausanne, Switzerland, in July 2024.

A Stellar Bash in Stunning Stellenbosch

I n 2007, the American Association of Wine Economists (AAWE) held its first annual meeting in Trier, Germany, followed by Portland the next year, establishing the tradition of alternating between foreign and domestic sites for its yearly gathering. The seventh conference was held for the first time in the Southern Hemisphere in early winter, from June 26 to 29, 2013, in Stellenbosch, South Africa, at Spier, a wine farm founded in 1692.

Despite the name of the society, one hundred thirty individuals from at least five continents registered to attend. These included a large contingency of academics, a few from government and industry, and a handful of consultants. The attendees represented a wide range of disciplines including agriculture, viticulture, business and management, marketing, statistics, mathematics, and even economics.

The Wednesday night welcome reception began with a brief address by Western Cape province Premier Helen Zille. Her thoughtful comments focused on both the blessing and the curse of alcohol in her country. While the over three-hundred-year-old wine and spirits business is taking off and creating new opportunities, alcoholism remains a serious problem, stemming in part from the legacy of paying workers with bottles instead of in rand. The system, known as tot or dop, was widespread on the wine farms of the Western Cape until it was finally banned just ten years ago.

I was part of a group who organized a blind tasting of South African wines that took place during the reception. The purpose was to illustrate two ways of quantifying and aggregating tasters' preferences to arrive at a societal outcome. Also, we wished to test the reliability of tasters who were selected randomly from among the attendees. Fifteen tasted eight samples of sauvignon blanc while another fifteen rated eight pours of pinotage, the South African cross between pinot noir and hermitage, known to us as cinsault. The results of the tasting, which are revealed below, were reported at a plenary session on Friday morning.

Twenty sessions, each with four-to-five presentations, filled two days. The topics included determinants of demand; wine trade and international wine markets; sparkling wine, beer, and food; notes on wine tasting; wine, wine farms, and financial markets; wine, wine farms, and the environment; innovations in wine marketing; the demand for sustainability in wine buying decisions; regulating the industry; wine and the macro-economy; country studies; production and efficiency; wine and the environment; and wine value chains.

My presentation, "Three Vignettes about Wine Tastings and Competitions," led off a special session on wine tasting organized by Elliott Morss, an economist and consultant based in Lenox, Massachusetts. The first vignette described an analysis of data from the 2010 Oregon Wine Awards that quantified the distortion in the outcome caused by easy graders.

The virtues of the California State Fair Commercial Wine Competition numerical rating scale were the subject of the second vignette. The third vignette explained what additional information can be wrung from the Borda scores of the societal outcome using as an example the Virginia versus Southern Oregon tasting results reported in "Vindicating Thomas" (see p. 298). The key message of the talk was that it is critically important to know the goals or objectives of a tasting or competition to select the most appropriate way of scoring.

Of particular interest to Oregonians was a talk by Ömer Gökçekuş, professor of international economics and development in the School of Diplomacy and International Relations at Seton Hall University in South Orange, New Jersey. His presentation, coauthored with master's candidate Clare Finnegan, was titled "Classification and re-classification: Oregon's Willamette Valley AVA [American Viticultural Area] and its new sub-AVAs."

The researchers "hypothesize that the establishment of sub-AVAs in the Willamette Valley was part of a dynamic process; it was an act of reclassification by 'better' wineries to distinguish themselves from 'lesser' wineries and ultimately collect a higher regional reputation premium." The pair calculated the price-to-quality ratios of wines produced in each of the six sub-AVAs before and after their establishment using ratings of pinot noir from the *Wine Spectator* as the measure of quality.

They concluded: "Before reclassification: sub-AVAs's quality is higher but prices [are] not. After reclassification: both price and quality [are] higher. Quality gap is not widening but the price/quality gap is. Sub-AVAs are collecting a much higher regional reputation premium."

At the plenary session on Friday morning, I led off the presentation of the results of the Wednesday evening tasting. The numerical ratings, based on a scale from 50 (poor) to 100 (excellent), recorded by each taster were averaged across the fifteen for each of the flights. The ratings were also converted to rankings.

Borda scores were determined by assigning seven points to the top-ranked wine in each flight, six to the second-ranked, and so on down to zero for the last place sample, with the scores for ties determined by the average of the Borda scores assigned to the rankings the group occupies (see "Borda is Better" on p. 16 for more information).

The sum of the Borda scores for each sample determined the consensus ranking of the eight. The range of average ratings for the sauvignon blanc was 76.9 to 82.3 (Fair/Mediocre to Good/Above Average) while the range of the cumulative Borda scores was 39.5 to 70.5. For the pinotage, the ranges were 72.1 to 78.2 (Fair/Mediocre) and 35 to 66.5. In neither flight were there any runaway favorites.

To gauge the reliability of the tasters, one of the wines in each flight was poured from the same bottle three times, so there were six different wines and two replicates sampled. This method was first used in 2003 at the California State Fair Commercial Wine Competition at the suggestion of Robert Hodgson, a California winery owner and retired oceanographer, who has gained international notice for his analysis of the consistency of awarding medals at competitions as well as for his study of wine judges' consistency.

Bob, who also presented at Morss's special session, analyzed the results of this tasting. Based on the outcomes of the single flights, he determined that, as a whole, the group of tasters was reasonably consistent in rating the three samples of the same wine similarly. Ideally, however, Bob recommends that a judge's consistency should be measured across four flights to attain a higher confidence level.

Knowing that all this talk made the attendees thirsty, a large and varied selection of mostly agreeable South African varietals and blends were served at the reception and with the lunches and dinners. Standouts included a 2009 Mourvèdre from Spice Route and a 2013 Sauvignon Blanc from Fleur du Cap. An untypical and unremarkable pinot noir was poured at one of the dinners.

On Saturday, the attendees went on one of three tours. I visited Simonsig, the first to produce Méthode Champenoise sparkling wine in South Africa; Kanonkop, my favorite—the still youthful 2002 pinotage was a treat, (as with other examples I tasted during my visit, it was happily more refined than rustic, more "pinot" than "-age"); and the gorgeously sited Glenelly, purchased ten years ago by the owner of Château Pichon Longueville Comtesse de Lalande, a second-growth bordeaux. The three tours came together for a late lunch at Fairview where a 2012 viognier and a 2011 durif (aka petite sirah) were particularly enjoyable.

The site for the eighth annual meeting was announced. Next year, in contrast to requiring three flights totaling almost 13,000 miles to return home from South Africa, I'll be making the half-day drive to Walla Walla, Washington.

Elliott Morss died in 2021. I miss his quirky perspective as evidenced by his staunch advocacy of the superiority of box wines.

Not mentioned in this story is the heated attack on using the Borda count in wine competitions launched by Dominic Cicchetti, a renowned biostatistician at Yale, an active member of the AAWE, and a prolific contributor to the literature of wine economics. His objection was that converting point scores to rankings loses information about the relative quality of the wines.

"If the only question is the relative ranking of 2 or more wines, specifically, which of the wines does one most prefer, quite irrespective of the perceived quality of the wine—as say, Poor/Unacceptable; Fair/Mediocre; Good/Above Average; or Excellent to Superior—then a ranking alone might suffice. This said, the distinct limitation of using ranks instead of [point] scores is the fact that very different wine [point] scores can and will receive the same relative rank ordering." (Cicchetti, Domenic, "Blind Tasting of South African Wines: A Tale of Two Methodologies," AAWE Working Paper No. 164, August 2014).

My concern with the reliance on point scores to arrive at a consensus ranking is that easy graders will have a disproportionate influence on the outcome whereas Borda gives each judge equal influence. Furthermore, the differences in Borda scores (as distinguished from point scores) indicate strength of preference, thus providing a measure of relative quality without distortion.

Sadly, Dom died in 2019. I miss his intelligence and collegiality.

Wine Economists Powwow
in Walla Walla

The eighth annual conference of the American Association of Wine Economists (AAWE) held from June 22 to 25, 2014, in Walla Walla, Washington, was something of a homecoming for the organization. Whitman College, the incubator of the society and its increasingly influential journal in the early years of this century, hosted a reception and two days of meetings and lunches on its charming campus, lavishing attention on the attendees as might a *kvelling bubbe* on her precocious grandchild.

We enjoyed the serenity of summer's first full day at an afternoon welcome reception at Whitman's Baker Faculty Center. Outside, several wineries poured white and red Bordeaux blends and Rhône varieties to accompany an appetizing buffet. The crisp 2013 Cadaretta SBS, a sauvignon blanc-sémillon blend, was a clear standout.

Registrants numbered one hundred ten and came from five continents to listen to seventy papers spread across fifteen sessions. In them, grape geeks of various analytical backgrounds described applications of the scientific method to a host of questions relating to wine, beer, whiskey, and even coffee.

Robert Hodgson, who famously questioned the value of wine competitions in a series of studies, gave a paper entitled "The unimportance of terroir" in which he demonstrated that AVA (American Viticultural Area) designated wines from Sonoma scored no better than county-designated wines but do command higher prices more frequently.

Robin Goldstein of Fearless Critic Media led off a session on "Quality and Experts" with an amusing yet thought-provoking presentation entitled "Do more expensive things generally taste worse?" A series of experiments he and others conducted indicate that the answer is yes, leading him to coin the phrase "inverse-pleasure effect."

While there is almost no disagreement among the attendees about the shortcomings of rating wines and vintages on a numerical scale, such scores are regularly used as proxies for quality in the absence of anything better. One such example was a presentation by Ömer Gökçekus and Clare Finnegan of Seton Hall University called "Lumping and splitting in expert ratings' effect on wine prices."

Lumping refers to the "[p]ractice of grouping similar things 'together in a single mental cluster'" while splitting is the "[m]ental action of 'perceiving 'different' clusters as separate from one another.'" Using data from *Wine Spectator* on 2,050 Willamette Valley pinot noirs, the authors demonstrated that while experts might subdivide ratings into several mental clusters, thus treating ratings as continuous, regular folks only consider two clusters—wines rated 90 and above and all the rest. The effect of a wine being placed in the higher mental cluster resulted in a significantly higher price.

My talk, entitled "Information Lost: The Unbearable Lightness of Vintage Charts," updated consensus rankings of Oregon pinot noir vintages that appeared in "Consensus Rankings of Oregon Pinot Noir Vintages: A Clash of Sensibilities," (see p. 54 for details); presented aggregated rankings for Washington State big red wines and elaborated on the inadequacies of both the constituent and consensus charts.

The first plenary session featured noted geologist Kevin Pogue of Whitman College who gave a thorough and thoroughly engaging overview of "The Terroirs of the Walla Walla Valley AVA."

At the second plenary, a distinguished panel examined "Regulation in the US Wine Industry." AAWE President and Princeton University Economist Orley Ashenfelter, Paul Beveridge of Family Wineries of Washington State, San Francisco attorney John Hinman, and Allen Shoup of Long Shadows shared insights from a range of perspectives on the continued effects of Prohibition and the battles to bring America's commerce in alcohol into a more practical place.

To balance the more serious economics discussions, lunches and dinners served as settings for sampling a wide range of local wines, thereby associating the two as the group's name suggests. Shoup presided over a banquet in the winery's barrel room. Among the impressive array of bottles available to sample were a 2007 Eleganté Cellar Cabernet Sauvignon from Walla Walla and Long Shadows' 2011 Sequel Syrah from the Columbia Valley.

A dinner at the celebrated Whitehouse Crawford restaurant in town featured delicious local cuisine and a number of seasonally and culinarily appropriate rosés, along with a few whites and reds. We particularly enjoyed the 2013 Amavi Cellars Cabernet Franc Rosé.

The best, however, was saved for the last day. As is the tradition, attendees abandoned meeting rooms and boarded buses to tour area vineyards and wineries. We rode a short distance south into Oregon where Dr. Pogue took us on an insider's tour of vineyards in the Milton-Freewater area.

Our first stop was Cayuse's Cailloux Vineyard planted in what became Oregon's eighteenth AVA in 2015, The Rocks District of Milton-Freewater. While teetering on the smooth volcanic stones, we learned that Dr. Pogue, who prepared the AVA petition, deliberately delineated its boundaries so that 97 percent of the soil is of one type, thereby establishing a single terroir. A subsequent stop showed us how starkly different the soil types are on each side of the boundary.

Returning to the Washington side, we tasted at Amavi Cellars, lunched at Basel Cellars, visited a winery and brewery incubator in the airport area, and toured the Center for Enology & Viticulture at

Walla Walla Community College, a major sponsor of the conference, sampling impressive efforts at each.

While my palate now decidedly favors the complexity and finesse of pinot noir, it rediscovered the pleasures of Washington State big reds. Stops at four tasting rooms in the downtown area allowed me to diversify my collection with bottles, many of which are from 2011. Though that vintage came in 14 out of 21 in the consensus ranking, I found the wines more elegant and better balanced than many I had tasted in the past from higher-ranked years. This year, one word most aptly summarizes our experience in Walla Walla: sweet.

A delightful surprise was reconnecting with Lynne Stokes, an accomplished statistician who was on the faculty with me at Vanderbilt University. We later collaborated on a study of whether Willamette Valley pinot noir constituted a cultural good (see p. 30).

The following year, the conference was held in Mendoza, Argentina.

Don't Cry for Us

Wine Economists Meet in Mendoza

Returning to the Southern Hemisphere for the second time since 2013, members of the American Association of Wine Economists (AAWE) convened in Argentina for the ninth annual conference held from May 26 to 30, 2015, and hosted by the National University of Cuyo (UNCUYO), Mendoza. An afternoon reception the first day was followed by two days of presentations and two days of wine tours, one more than in the past.

The Welcome Reception was held at the Park Hyatt Mendoza. As we mingled with colleagues whom we hadn't seen since the last meeting and met new ones, several bodegas poured local wines that accompanied appetizers including Argentine beef. A chorus from UNCUYO serenaded us with wine-related songs.

Serious business began early the next morning and continued through the following day. The largest number of presentations ever—one hundred forty—were accepted. Over one hundred were presented by speakers from six continents and represented a remarkably broad range of disciplines including economics, statistics, marketing, viticulture, and planning. Plenary sessions covered productivity, research and development, and history. In addition, five blocks of four parallel sessions each covered climate change and sustainability, reputation and region, tourism and new products, regulation and politics, demand, organization, prices, production, marketing, tasting, Argentina, non-wine, health, global and trade, and miscellaneous topics.

At a miscellaneous topics session I chaired, I presented a talk with a most nerdy title: "Minimum Percent Error – Zero Percent Bias Regression for Wine Economists." The purpose was to introduce an alternative way to fit formulas to data that was developed by cost analysts to build models to estimate cost, schedule, or technical parameters. The intent was to transfer this approach from one community to another that had similar needs for a wider range of predictive tools.

Dan Moscovici of Stockton University led off the session on climate change and sustainability with a global survey of practices. He noted that New Zealand was the first to emphasize sustainability in 1994 and that the United States does not have a standard platform but, rather, a potpourri of accreditations including Oregon Certified Sustainable Wine. While the International Organization of Vine and Wine (OIV) offers five criteria, his goal is to define a wine economics-based standard for sustainability as an international best practice that allows one to "be sustainable without spending a lot of money."

In the same session, Maryam Hariri, a planner at New York University, discussed her research into the relationship between temperature and labor supply. She looked at data from wineries in New Jersey and New York to gain an "understanding of how different variables (for example, size, management, and financial structure of winery) influence a winery's capacity to gain access to the labor resources needed to ensure timely response to extreme weather events."

The session on tasting featured presentations by several well-known researchers including Robert Hodgson of Fieldbrook Winery and Robin Goldstein of the University of California, Davis. Hodgson tested a claim that experts prefer more expensive wine and concluded, based on "several thousand observations of expert wine judges covering several years of major US wine competitions," that they do slightly but not significantly.

Observing the poor agreement among judges in wine competitions, Jing Cao and Lynne Stokes of Southern Methodist University examined whether this variability is predictable. Using the same database as Hodgson, they concluded that varietal type and flights of between ten and forty wines do not affect judges' agreement whereas flights of less than ten wines "are associated with lower agreement among judges." Also, "judges have consistent (and slightly better) agreement as the process of wine evaluation proceeds, which indicates that fatigue does not affect their performance within the range investigated."

Goldstein presented a progress report on his attempts to explain the Price-Quality Inversion (PQI) his experiments involving blind tastings consistently demonstrated for various beverages and foods. A particular example of PQI is the preference for the least expensive wine over the most expensive, which is likely the result of the most expensive wine being tasted while still too young. He hypothesized "that wine expertise shapes individuals' sensory perceptions, giving rise to 'acquired tastes' for the types of expensive wines that most consumers start out disliking."

Tapas lunches, which invariably included empanadas, were served in a tent along with wines from various producers and featured among several other varieties, malbec, the grape Mendoza is best known for. While many could use more time in the bottle, one example that never saw oak provided a refreshing accompaniment. The most memorable wine, however, was the ethereal 2007 Bodega Benegas Cabernet Franc Benegas Lynch Libertad Vineyards from one hundred twenty-year-old vines.

Following each day of sessions, we went by bus to different places for dinner. The first was Bodega Septima, a striking facility that we, unfortunately, couldn't see because of mid-autumn's early sunset. The entertainment included tango dancers. The next night we went to Divina Marga, a multi-use facility, where we dined on indigenous Argentine food while four couples performed traditional dances.

On the first day of wine touring, we visited the renowned Catena Zapata Winery housed in a Mayan-inspired building surrounded by the Adrianna Vineyard, a source of some of the most celebrated malbecs and cabernet sauvignons in the area. We sampled two very young 2010 bottlings, the 100-percent Malbec and a surprisingly plush but nevertheless underdeveloped Nicholas Catena Zapata blend of cabernet sauvignon and malbec.

Next, we were bussed to Salentein Winery for lunch, which was served with a 2014 Sauvignon Blanc Reserve and 2013 Malbec Reserve. The elegant facility blended handsomely into the desert landscape framed by the snow-tipped Andes.

On the second day of touring, we first visited Casarena Winery where we tasted a wide range of wines including a 2012 Ramanegra Reserva Pinot Noir. We then lunched and toured at Familia Zuccardi Winery and Olive Oil Farm. We were treated to a remarkable selection of vegetables and meats grilled on a large parrilla accompanied by three Santa Julia wines: a 2014 Chardonnay, a 2013 Reserva Malbec, and a 2014 Tardío dessert wine. On each table were four different bottles of the olive oil made on the premises, including one that was recently pressed and unfiltered, which made particularly tasty bread dips.

Of the four meetings I have attended, this one in particular seemed to pinpoint some opportunities for significant international collaboration. In addition to the challenge set forth by Moscovici, the compilation of economic histories of wine regions around the world was identified as an undeveloped area. The confluence of graybeard wisdom, midcareer drive, big data, and analysis tools should be sufficient to ensure that future gatherings will have much to talk about.

Prior to the meeting, we visited Uruguay and toured two wineries with a private guide. Along with various versions of tannat, a teinturier grape from the Madiran region of France that has found a most congenial home in the tiny South American country, we enjoyed other varieties that do well there. After the meeting, we flew to Santiago, Chile, and had private tours of six wineries over two days. If I had to pick one country I would like to return to, it is Chile.

Unfortunately, I missed the Bordeaux meeting because we had been to Europe earlier in the summer of 2016 for a tour of Champagne, and the schedules didn't link up as they had in 2013 when we went from a Burgundy tour to Stellenbosch for the meeting there.

In the Garden of the
Wine Economists

Orto Botanico di Padova Provides Fertile
Ground for a Burgeoning Field

For its eleventh annual meeting, the American Association of Wine Economists (AAWE) returned to Europe for the second consecutive year, moving from Bordeaux, France, in 2016 to the ancient university town of Padua (Padova, to the locals) in Northern Italy. Some two hundred researchers and their spouses from around the world registered for the conference, which was held from June 28 to July 2, 2017. Our host was the Università degli Studi di Padova (University of Padova), founded in 1222.

Those arriving early on the first day were treated to a guided tour of the Old University Building. Later in the day, the welcome reception was held in the Paladin Room of the Palazzo Moroni (Town Hall). Two days of presentations and two days of wine tours followed.

On the second day, we gathered in the Orto Botanico. Dating from the mid-1500s and now home to some 7,000 plant species, it is "the oldest university garden in the world to have retained its original location and layout," according to its website (http://www.unipd.it/

en/university/cultural-heritage-0/botanical-garden). The technical meetings began with a plenary session that included presentations by three AAWE scholarship winners. The topics covered consumer reaction to brewery acquisitions, the impact of the European migrant crisis on vineyard productivity in Southern Italy, and a new method for assessing and communicating on wine-cheese pairings.

Next was a roundtable discussion on the future of Veneto Quality Wine featuring six members of that region's wine industry. The Veneto region, of which Padua is a part, grows almost 20 percent of Italian wine. Prosecco, a bulk-processed sparkling wine made primarily from the simple glera grape, is the most important product since it has a growing following in North America and Europe, though local consumption is down. Strategies to increase exports to non-European Union countries were discussed along with efforts to increase quality.

Of the 164 presentations accepted, about one hundred forty were given in twenty sessions, with three or four held simultaneously. The topics included trade; profiling wine consumers; regional wine identity; wine business; producing grapes; sustainability and consumers; tasting, quality and experiments; wine history; determinants of wine prices; wine tourism; growing grapes; wine experts and wine tasting; wine regions; quality, value, regulation; production, sustainability; global warming; and signaling wine quality.

Among the highlights of the presentations I attended were two studies that suggest Italian consumers were willing to pay more for organic or other socially responsibly produced wine. Denton Marks of the University of Wisconsin-Whitewater channeled Daniel Kahneman, a Nobel laureate who is a pioneer of behavioral economics and author of *Thinking Fast and Slow*, in a talk entitled "How Do We Decide About Wine? Fast, Slow, and Otherwise."

Christopher Bitter of Vintage Economics gave a presentation entitled "Wine Competitions: Reevaluating the Gold Standard" which concluded, "that competitions vary substantially in their level of generosity and that most have become more generous in recent years."

Don Cyr, Lester Kwong, and Ling Sun of Brock University in Ontario, Canada, offered "An exploration of the relationship between Robert Parker and Neal Martin *en primeur* wine ratings." I was amused to see an application of a relatively obscure statistical entity that I had run into in a former career called a copula (yes, it is from the same root), which is a multivariate probability distribution, to compare scores.

In 2008, Robert Hodgson, a professor emeritus at Humboldt State University in Arcata, California, published a frequently cited study debunking the reliability of wine judges in the California State Fair Commercial Wine Competition. In "2008 Revisited," he added the results of the 2009 to 2012 competitions to those from 2005 to 2008 and confirmed his original conclusion. "Based on ...criteria [that he established], less than 6 percent of the wine judges yielded an acceptable rating."

In the session entitled "Signaling Wine Quality," I presented "Toward Valuing Willamette Valley Pinot Noir as a Cultural Good" coauthored with Lynne Stokes of Southern Methodist University. We addressed the question of whether the price of Willamette Valley pinot noir reflects its value as a cultural good as opposed to just another commodity or agricultural product.

A survey was administered to visitors to four tasting rooms. Each respondent tasted two pinots and completed the survey form for each. Regression models established the association between willingness to pay (WTP) and the various responses, demographic information, and attributes of the wines with aesthetic value ("I find this wine beautiful") most strongly associated with WTP, with an increase of about $9 for tasters who endorsed this statement compared to those who didn't.

Like all practitioners of the dismal science, wine economists are two-handed. But unlike those who frustrated Harry Truman, they are almost always holding a glass in one of them. Meals and tours provided ample opportunity to do so. Buffet lunches served on the terrace of the Botanical Garden were accompanied by local wines. Most intriguing were the experimental hybrids made by Vivai Cooperativi

Rauscedo from sauvignon blanc, tocai friulano, and pinot noir crossed with numbered but otherwise unnamed *vinifera* varieties. After the first day, we were taken to Abbazia di Praglia in Teolo for a tour and dinner. The Gala Dinner marking the end of the technical sessions was held at the historic Restaurant Pedrocchi in downtown Padua.

The first day of wine tours focused on the Conegliano Valdobbiadene region, the hub of the Prosecco Superiore wine industry. We visited the Enological College, toured the labs and winery, had lunch at the school accompanied by Prosecco, of course, and then split up into four groups to visit a pair of wineries. Il Colle, a relatively new and modern producer, and Villa Sandi, a larger winery, were on my itinerary. The bus ride between the two showcased stunning views of the steep—up to forty-five degrees—vineyard-covered hills.

The next day we went to Venice for a boat tour around the lagoon. The first stop was at San Erasmo Island, home of Orto, makers of an unusual white wine that is "an assembly of antique Italian grape varieties with a dominance of the Istrien Malvoisie." We tasted the 2015 and 2008 vintages, the latter displaying some slight oxidation that I found pleasant but was off-putting to at least one of my colleagues. The island of Torcello was where we had lunch and a postprandial stroll to an old vineyard. Piazza San Marco was our final stop on the island of Venice.

AAWE, now in late adolescence, continues to blossom with steady and growing worldwide participation in its conferences and a journal that is regularly cited not only in academic publications but also in the popular press. As it evolves its mission, the organization has embraced an impressive range of disciplines that make these meetings endlessly stimulating. For this conference, the Orto Botanico di Padova was a particularly appropriate site since it is a successful demonstration of enduring diversity.

For the twelth annual conference, the AAWE returned to American shores for the first time since 2014. Cornell University in Ithaca, New York, hosted the meeting from June 10 to 14, 2018.

Wine Economists Convene "Far above Cayuga's Waters"

The twelfth annual meeting of the American Association of Wine Economists (AAWE) was hosted in Ithaca, New York, by Cornell SC Johnson College of Business and Cornell College of Agriculture and Life Sciences with support from the New York Wine and Grape Foundation. Some one hundred twenty researchers and their guests from around the world registered for the conference, held from June 10 to 14, 2018, on the Cornell University campus "far above Cayuga's water," as its alma mater observes, in the Finger Lakes district.

The Finger Lakes American Viticultural Area (AVA) in upstate New York, created in 1982, is the largest in the state. It now houses about two hundred forty wineries and includes the Cayuga Lake and Seneca Lake AVAs.

Cayuga Lake is the longest of the Finger Lakes while Seneca Lake is the largest and deepest. The Finger Lakes take their name from the eleven lakes deeply clawed out by glaciers beginning around two million years ago and splayed like fingers on hands. This cool climate region's principal *vinifera* varieties are riesling, cabernet franc, chardonnay, and pinot noir. French-American hybrids and native varieties still dominate production, however.

In the late afternoon of the first day, the welcome reception was held at the Herbert F. Johnson Museum of Art on the campus of Cornell University with a great view of Cayuga Lake. As has become the tradition, the next two days of presentations were followed by two days of tours.

In all, seventy-one presentations were given in five technical sessions, with two or three held simultaneously. The topics included tourism; market, history and future; prices, auctions, and investing; organization; sensory and experiments; income, politics, sustainability; regional markets; geography, clusters, innovation; consumers and marketing; experts and tasting; trade and global; and consumer preferences. In addition, there were three plenary sessions including one on cannabis for the first time in the organization's history.

Among the highlights of the presentations in the plenary sessions was the overview of wine from the Finger Lakes by John Winthrop Haeger, author of *Riesling Rediscovered Bold, Bright, and Dry,* and two books on pinot noir.

Qian Janice Wang and Domen Prešern, AAWE Scholarship recipients at the University of Oxford and members of its Blind Tasting Society, gave an excellent presentation entitled "Does Blind Tasting Work? Investigating the Impact of Training on Blind Tasting Accuracy and Wine Preference" based on an analysis of results of training sessions conducted by the society.

The first-ever session on Cannabis and Opioid Economics was the last of plenaries. Among the talks were "Opioids and alcohol: Substitutes or complements, and the implications" by AAWE President Orley Ashenfelter of Princeton University and "Is cannabis like wine? The user experience offered by some popular cannabis strains" by James Evans of University of Laval in Canada, which recently legalized marijuana.

The session concluded with Robin Goldstein of UC Davis offering a rendition of "Regulation's On," a topic-appropriate takeoff of Bob Marley's "Redemption Song." ("Economists, yes, they rob I…")

Among the presentations in the technical sessions was "Why and how do we blend? Comparing different trends for coffee and wine" by Morton Scholer, author of *Coffee and Wine: Two Worlds Compared*, which was released in 2018. He noted that the coffee industry has embraced the idea of *terroir* from the wine world.

A "Behavioral perspective on fine wine pricing" by Polish researchers Paweł Olekey and Marcin Czupryna, both of Cracow University of Economics, and Michal Jakubczyk of SGH Warsaw School of Economics, considered the impact of atypical features such as non-standard bottle size and packaging defects.

Joseph Breeden and Sisi Liang of auctionforecast.com addressed the challenge of "Forecasting auction prices of illiquid wines" by comparing the results of several models. Based on a wine tasting conducted during a celebration of her university's fiftieth anniversary, Britta Niklas of Ruhr-University Bochum in Germany reported mixed results in her talk entitled "Can music change the taste of wine?"

In the session on "Experts & Tasting," I gave a presentation entitled "How to Decide How to Decide" that addressed the confusion surrounding the selection of the best way to aggregate rank-ordered preferences. In the world of wine economics, the issue arises when tasters provide rankings of wines that must be combined to yield a single ranking. It has also come up when a consensus ranking of vintages was being sought (see p. 54). The presentation served as a reminder that the Borda count (see p. 16) has been proven to uniquely satisfy four rational properties and should be used exclusively to determine the consensus ranking.

Other talks in the same session included "Wine, women, and men: Large sample results" by Jeffrey Bodington, who concluded that there was no significant difference between the scores assigned to wines by male and female judges at the 2016 Wines of Portugal Challenge. This confirmed his earlier study based on smaller sample sizes.

One of the Best Paper Awards went to "Fine Water: A Blind Taste Test" by Kevin Capehart of California State University, Fresno, and Elena Berg of the American University of Paris. Applying methods used for other beverages to conduct and analyze the results of a double-blind tasting of bottled water, the authors concluded that trained subjects performed slightly better than chance at distinguishing between different brands of water and that subjects were no better than chance at distinguishing tap water from bottled. (No, the tasters didn't cleanse their palates between samples with sips of wine. I asked.)

Local wines from several producers were poured at the welcome reception and accompanied the lunches and dinners. In addition, a wine from Castello Banfi, an Italian producer and Cornell donor, and a barrel sample of Orley Ashenfelter's 2017 cabernet sauvignon, grown in his vineyard in New Jersey, were offered. On the third evening, after the completion of the technical portion of the conference, we set sail on the *Seneca Legacy* for the Gala Dinner Cruise on Seneca Lake.

The first day of tours began at the deservedly acclaimed Ravines Wine Cellars on Seneca Lake. After being greeted with a truly impressive 2008 Blanc de Blanc sparkling wine, we visited White Springs Vineyard next to the winery and the white and red wine production facilities, tasting at each stop.

Lunch was prepared by Lisa Hallgren, co-owner with her husband and winemaker, Morten, and served with a selection of lovely Ravines wines. I particularly enjoyed the elegant and subtly complex 2016 White Springs Vineyard Riesling and the 2014 Maximilien, an approachable and savory blend of cabernet sauvignon, cabernet franc, and merlot, named for the owners' mathematician son (Yes!).

After lunch, we divided into three groups and headed to Keuka Lake, West Seneca Lake, or East Seneca Lake. I chose Keuka Lake, home of the pioneering winery established by Dr. Konstantin Frank in 1962. Frank was the first to successfully grow and market *Vitis vinifera* wine grapes in the Eastern United States.

We were welcomed with a glass of nonvintage Célèbre, a sparkling riesling made by the Crémant method. Konstantin's grandson and winery president, Frederick Frank, spoke to us about the history and vineyards, which are some of the oldest in the country, and the winery. His daughter, Meaghan Frank, the general manager, then led us through a tasting of nine sparkling and still wines.

Standouts were the refreshing 2017 Rkatsiteli, redolent of grapefruit; the beautifully balanced 2016 Eugenia Dry Riesling, named for Konstantin's wife; and the rare and rustic 2015 saperavi, a red grape with red pulp. Rkatsiteli and saperavi originated in the Republic of Georgia.

The next day we went to two cideries near Cayuga Lake. Finger Lakes Cider House treated us to a tasting of both still and sparkling Kite & String brand ciders, one of which included unfermented riesling juice and marechal foch red wine. Black Diamond Cidery, owned by retired Cornell professor, Ian Merwin, and his wife, Jackie, is a tiny facility that manages to make a wide variety of ciders from about one hundred fifty varieties of apples all grown on their farm. Porter's Pommeau, a dessert cider made with the addition of eau de vie in the style of a port, was particularly intriguing.

Still very much a boutique professional society, AAWE enters its teen years continuing to seek broader relevance by comingling its core and namesake interest in wine with other liquids including beer, spirits, coffee, and even water. Drugs, both legal and "should-bes," ("…Just turn-a-cannabis-a-to wine…") have now found a place on the agenda.

Here, I spill over a disagreement over the best method for aggregating the rankings of vintages by several critics that first appeared in an academic journal into a popular wine publication, the Oregon Wine Press. *A mistake caused by mistranscribing points assigned to vintages by* Wine Spectator *is corrected in Table 2.*

Consensus Rankings of Oregon Pinot Noir Vintages: A Clash of Sensibilities

I am of two minds: one oenophilic, the other mathematical. Since the former generally occupies the right side of my brain while the latter the left, most of the time the two stay out of each other's way. When they do meet, the outcome can be reasonable. But trouble can happen when passions collide.

Such is the case when my inner oenophile is confronted with scores and ratings, which it knows are naïve gross oversimplifications, while the mathematician can't resist anything quantified and adheres to the belief that when life gives you numbers, write an article. In the case of this piece, the mathematician initially prevailed. I couldn't help myself. An article recently appeared that prompted me to react.

Three members of the Faculty of Engineering at the University of Porto in Portugal—José Borges, António C. Real, and J. Sarsfield Cabral—and Southern Oregon University's Gregory V. Jones published a paper in the *Journal of Wine Economics* in 2012 entitled "A New Method to Obtain a Consensus Ranking of a Region's Vintages' Quality." The following motivation for the study is given:

> Understanding vintage quality variability and its influences are important in the economic sustainability of producers, consumer purchasing decisions, investor portfolio holdings, and researchers examining the myriad drivers of quality. However, the process of finding an adequate measure of vintage quality is a challenging task due to the paucity of information and the inherent subjectivity in assessing quality.

The authors assembled vintage charts "by internationally recognized critics, magazines, or organizations" for three wine regions—Piedmont, Burgundy, and Champagne—and converted the numerical ratings, which are given on various scales, typically five or 100 points, to rankings. The individual rankings were aggregated to arrive at a consensus ranking for each region.

Without going into too much arcana, I took issue with the method used to combine the rankings, which was advocated by and named for the French mathematician, Marie Jean Antoine Nicolas de Caritat, Marquis de Condorcet, since it has already been established that the Borda count uniquely satisfies a few highly desirable properties while using all of the information in each ranking to arrive at the consensus (see "Borda is Better" on p. 16 for more information).

Condorcet does not since it only considers pairwise contests between all the alternatives (in this case, vintages) and declares as winner the one that beats all others the most times. This obvious infraction of what should be regarded as "settled mathematics" became a case for the "Borda Patrol" and resulted in my publishing a comment on the paper in the next issue of the *Journal of Wine Economics,* which refutes the use of Condorcet and presents alternative consensus rankings based on Borda.

In the same issue, the authors of the original paper offered a refutation, which involves appealing to the fact that Condorcet satisfies a property that canonizes discarding part of voters' preferences. My only response is that folks are entitled to their own opinions but not their own mathematics.

Rather than leaving an opening for the propagation of a flawed approach to aggregating the rankings, I set aside, for the moment, my objections to assigning numerical ratings to vintages and used Borda to prepare consensus rankings of Oregon pinot noir vintages based on several sources. I offer two.

The first is the consensus ranking of vintages from 1997 to 2007 developed from seven charts prepared by *Decanter*, Decanted Wines, DeLong, Sorkin (IntoWine), the *Wine Advocate*, the *Wine Enthusiast*, and the *Wine Spectator*. The ratings given each vintage in each chart were converted to rankings, with ties permitted. Each ranking was converted to a Borda score as follows: the top-ranked vintage received a score of 10 (one less than the number of vintages considered); the second, 9; and so on down to the bottom-ranked vintage which received a score of 0.

In the case of ties, each vintage received the average of the Borda scores assigned to the rankings the group occupies. So, for example, if three vintages out of eleven are tied and occupy the third through fifth positions, each would get a Borda score of $(8+7+6)/3 = 7$. Note that ties are more common for Decanter and DeLong since they use a scale of 5 to rate vintages. The Borda scores for each of the seven sources, the sum of the scores, and the consensus ranking are presented in Table 1. The higher the sum of the Borda scores, the higher the ranking. The consensus of the seven ranks was 1999 at the top and 1997 at the bottom of the eleven vintages, for which ratings from all were available.

Table 1. Consensus Ranking of Oregon Pinot Noir Vintages 1997 to 2007 Based on Seven Charts

Rank	Year	Wine Advocate	Wine Enthusiast	Decanter	De Long	Decanted Wines	Wine Spectator	Sorkin (IntoWine)	Sums of Borda Scores
1	1999	9.5	9.5	7	9.5	8	7.5	9	60
2	2002	9.5	3	7	9.5	10	10	10	59
3	2006	8	9.5	7	5	5	5.5	8	48
4	2004	3.5	7.5	7	5	5	9	3	40
5	1998	7	7.5	2	5	7	5.5	4	38
6	2001	1.5	4.5	7	5	9	4	6.5	37.5
7	2000	3.5	6	7	5	5	3	0	29.5
8	2005	1.5	4.5	2	5	3	7.5	5	28.5
9	2003	6	2	7	5	2	2	1.5	25.5
10	2007	0	0.5	2	0.5	1	0.5	6.5	11
11	1997	5	0.5	0	0.5	0	0.5	1.5	8

If we consider only the ratings in the vintage charts from the *Wine Advocate*, the *Wine Enthusiast*, and the *Wine Spectator*, we can compile a consensus ranking for vintages from 1991 to 2009. The process of applying Borda is the same except we assign a score of 18 (one less than the nineteen vintages considered) to the top-ranked in each chart. The Borda scores for each of the three sources, the sum of the scores, and the consensus ranking are presented in Table 2. The 2008 vintage is at the top, and the 1995 is at the bottom of the nineteen vintages for which ratings from all three sources were available.

But my inner oenophile recoils at what the math hath wrought. Had not the Scott Paul Audrey and the Archery Summit Red Hills Pinot Noirs from the next to the bottom-ranked 2007 vintage performed the "Dance of the Blessed Spirits" upon my palate and sent me to my sock drawer for replacements? On the other hand, weren't many of the bottles from the third-ranked 2006 that I sampled overly zaftig, boozy, unbalanced, and, sadly, inelegant?

What is going on, then? As with the selection of a method to aggregate the rankings, a critical factor in assessing the relative merits of vintage is the use of all information available. Distilling this information down to a single number leaves out too much and yields unsatisfactory results.

As Allen Meadows, the Burghound, observes, "Like life, Burgundy vintages are simply far too complicated to be captured by a numerical rating on a chart."

Mrs. Burghound, Erica Meadows, elaborates, "[Allen] has said that most vintages vary too greatly because of Burgundy's complex topography. Hail can easily hit one village and spare the one next to it, and as such, he believes vintage charts can actually do more harm than good. There are simply way too many variables that come into play, not the least of which are the extremely localized weather effects, individual terroirs, and the quality of the growers themselves. He feels vintage charts can do more to mislead consumers than inform them and that buying on the basis of vintage is dangerous because one producer may have done spectacularly well, either by luck or skill, even in a poor vintage, whereas someone less diligent, or less lucky, may have performed poorly in a great vintage."

And thus is the case with Oregon vintages, I might add. Therefore, while the mathematician did the best he could with what he had but over which he had no control, the outcomes are ultimately unsatisfying and reflect the oversimplification inherent in each of the constituent charts.

So, what is a consumer or investor or any one of the supposed beneficiaries of these aggregated rankings to do? Believe no one but yourself. Trust your own palate and no one else's unless you have calibrated yours to theirs. If you are a researcher seeking the determinants of quality, you would be well advised to take consensus rankings with a grain of salt, as well. Better to read a range of reviews of a representative but a sufficiently broad range of samples from the vintages of interest and to seek other metrics such as total growing degree days, rainfall, and length of growing season as better representations of a particular year. That might give the mathematician something a little more credible to work with.

Table 2. Consensus Ranking of Oregon Pinot Noir Vintages 1991 to 2009 Based on Three Charts

Rank	Year	Wine Advocate	Wine Enthusiast	Wine Spectator	Sums of Borda Scores
1	2008	18	15	18	51
2	1999	16	17.5	14.5	48
3	2006	14	17.5	12	43.5
4 tied	2002	16	6.5	17	39.5
4 tied	1998	12.5	15	12	39.5
6	1994	16	10	12	38
7	2004	6.5	15	16	37.5
8	1993	12.5	12	3.5	28
9	2005	4.5	8.5	14.5	27.5
10	2000	6.5	12	8.5	27
11 tied	2009	2.5	12	8.5	23
11 tied	2001	4.5	8.5	10	23
13	1992	10.5	6.5	5	22
14	2003	10.5	2.5	6.5	19.5
15	1991	8.5	4.5	3.5	16.5
16	1996	1	4.5	6.5	12
17	1997	8.5	0.5	1.5	10.5
18	2007	2.5	0.5	1.5	4.5
19	1995	0	2.5	0	2.5

Georgia for the Wine

Join the American Association of Wine Economists (AAWE) and see the world, I like to say. Since 2012, I have attended annual meetings in Stellenbosch, Mendoza, and Padua, as well as a few in the US. After two years of no gatherings, I jumped at the opportunity to visit the Republic of Georgia for the first time to attend the fourteenth annual conference from August 24 to 28, 2022, in its capital city, Tbilisi. A country whose tagline is "8000 Vintages" was an obvious and thirst-inspiring choice. Georgia's claim to be "The Cradle of Wine" rests on evidence of grape growing and wine production as early as the middle of the sixth millennium BCE. *The Wines of Georgia* (Infinite Ideas, 2020) by Lisa Granik MW is an authoritative reference for those interested in too much information.

Wine permeates Georgian life. There are numerous wine bars and shops throughout Tbilisi. Many residents in this country of slightly over four million, if you count the Russian-occupied territories—just under 3.7 million if you don't—maintain small vineyards and make their own wine, as has been done for centuries.

While there are at least four hundred Georgian grape varieties, only a handful are used to make wine. About 75 percent of the wine grapes grown are white, with rkatsiteli by far, the largest planting. Saperavi is the most abundant red. International varieties can also be found, including a small amount of pinot noir used to make dry and sparkling wine. (While Dr. Konstantin Frank Winery in the Finger Lakes of New York grows and bottles tasty rkatsiteli and saperavi, there are no reports of anyone in Oregon experimenting with these or other Georgian varieties.) Traditionally, wines were made in large clay containers called *qvevri*, known as *churi* in Western Georgia. This continues today, but modern techniques are also gaining ground and are now dominant. Kakheti, in the far eastern part of the country, is the source of 75 percent of wine grapes and 68 percent of wine.

With war raging on the other side of the Black Sea, instigated by Georgia's neighbor to the north, and lingering concerns about the pandemic, attendance was down. As Karl Storchmann of New York University who serves as Executive Director of AAWE, commented, many wives forbade their husbands from going. Fortunately, mine didn't, but decided not to go herself. So I joined the smaller-than-usual international group of academics, marketing folks, and others. We enjoyed the welcome reception the first evening, followed by two days of presentations at Tbilisi State University and two more of touring.

Since my wife stayed stateside, my son, Kenny, an oenophile in his own right, decided to tag along. When we converged, we hied over to "Wine Not?" a sweet little wine bar and shop a short walk from our hotel. Ana Berikashvili hosted us.

After we told her we would like one glass of white and one of red, she offered a taste. It pleased us, so we selected it as our white. Since she wanted an honest opinion, only after we approved did she tell us she was the winemaker. The wine was a mtsvane, another popular white variety, from her AnaBerika Winery in Kakheti.

The next day, we visited the closest of four locations of "8000 Vintages," a much-heralded shop with an overwhelming collection of Georgian wines racked floor to ceiling. We were assisted by a young woman who was studying winemaking. We picked up two authentic ice wines for Kenny and sampled rkatsiteli and saperavi in advance of a second visit I planned after the conference. When I did return, I let her pick out some bottles to add to my stash, further detail below.

On the second full day of presentations, I chaired the first session. In it, I summarized my analysis of the results of the 1980 wine tasting that led to the establishment of Domaine Drouhin Oregon.

"*Contre-degustation Olympiades du Vin* According to Borda" compares the rankings of mostly pinot noirs and mostly chardonnays obtained by averaging the score assigned by the judges with those using the more mathematically justifiable Borda count. While there were some differences overall, both methods ranked the 1975 South Block Reserve Pinot Noir from The Eyrie Vineyards second to the 1959 Chambolle-Musigny from Joseph Drouhin, which might be considered a ringer due to its age. In any case, this impressive showing convinced Robert Drouhin to open a winery in the Dundee Hills in 1987, the first Burgundian to do so.

One notable attendee was Robert Joseph, the British wine writer aka the Devil's Advocate, who among other roles is co-owner of Le Grand Noir wines and editor-at-large at Meininger's Wine Business International. His main duty seemed to be to keep it real, in the face of some very academic and arcane discussions, and to serve as an advocate for wine-drinking hoi polloi.

After my presentation, Joseph explained that, in Europe, Borda is used to select the "best of" category in wine competitions, something that isn't routinely done in the US. This, of course, led to my enumerating my objections to wine competitions, for which I have little regard. As a comparative wine snob and, worse, an opinionated nerdy analyst, I enjoyed the repartee. Later, in a side conversation, he emphasized that the focus of all wine-related discussions must be the consumer. Sage advice since, ultimately, that is what each wine person is.

Later that day, I gave my second presentation, "What Can I Still Afford to Drink?" This looked at the current availability and prices of the mostly French and German wines I tasted between 1969 and 1979, which were recorded in four notebooks. I also found the prices of more recent vintages that are the same age as those tasted in the 1970s. To establish the upper limit of affordability, I simply used the price of the most expensive bottle in my collection. I also compared current prices to the inflated prices, using the factors for wine created by the Bureau of Labor Statistics. Since these are intended for commodity wines, they resulted in much lower estimates of prices for the higher-end collectible bottles. This leaves open the question as to whether there might be a way to predict future prices of luxury wines, over time, based on an appropriate selection of factors.

To my surprise and delight, my second paper received one of three Christophe Baron Prizes for the Best Conference Presentation. Instead of a non-consumable certificate, we were awarded bottles of 2018 Cayuse Syrah from the Armada Vineyard. Much preferred.

One of the many joys of these conferences is the stream of local wines served during breaks and at lunches and dinners. There are simply too many to mention, but toward the end of the dinner at the Funicular Restaurant high above the city, we were poured a 2016 Shalauri Cellars Saperavi from a decanter. Despite being well on my way to satiation, it grabbed my attention. The story of how I ended up with a complementary bottle involves the persistence of a former freelance journalist from the UK who decided she would repay the kindnesses she had been shown by paying it forward to me. This bottle paired very well with a vegetarian chili.

On the first day of touring, we visited Ikalto Monastery and Ikalto Academy, where viticulture and enology were taught in the twelfth century. We sampled two dozen wines in Tsinandali and dined and tasted at Mosmieri, a winery founded by Joerg Matthies, a German industrialist, who picked out three wines for me to purchase.

On the second day, we visited Uplistsikhe Cave Town, one of the oldest urban settlements in Georgia, where I risked life and limb scampering on and over steep paths of varying widths and depths, trying to maintain a foothold while being buffeted by brisk winds. The reward was getting up close to the remains of an ancient settlement that included a winery. I'm still debating whether the risk was worth the reward.

We then went to Château Mukhrani, which was founded in 1878 by an heir to the royal family of Georgia. An excellent tour of the impressive facilities included both traditional and modern winemaking vessels as well as a distillery for making *cha cha*, the Georgian version of grappa. Then we tasted four wines before dinner. While I was sampling one of the bold reds that are a hallmark of Georgia, a couple approached me and enthusiastically asked if I thought it was like a Willamette Valley pinot noir. Not even close, I insisted.

The day I left, I returned to "8000 Vintages" to fill out a case. Many of the whites are made with long skin contact, which imparts a deeper color and draws more tannins, ensuring ageability. I look forward to opening the amber 2017 Rkatsiteli from Tchotiashvili Family Vineyards. The most expensive bottle, a 2015 saperavi from Badagoni, cost $38.40. Mercifully, the labels are in both English and Georgian. The incomprehensible script looks to this linguistically challenged observer like a collection of squiggles and artfully arranged picture hooks punctuated with tridents and the universal on/off sign.

Until recently, little of the wine produced made it to Western countries. After the collapse of the Soviet Union, which had been the single largest importer of Georgian wines, the emphasis began to shift from quantity to quality. Those seeking something completely different from the international varieties and pervasive winemaking styles would do well to look for bottles from Georgia.

Back to Stellenbosch in 2023

Though we enjoyed our first visit in 2013, we didn't expect that we would ever be returning to South Africa, given the distance and long travel time. From Portland, Oregon, the trip takes well over a day, including layover times and at least twenty hours in the air. But when the 15th Annual Meeting of the American Association of Wine Economists (AAWE) was scheduled for the Krystal Beach Hotel in Gordon's Bay, from June 28 to July 2, 2023, we knew we were going to go back.

To break up the trip, we stopped in London for a few days. This gave me the opportunity to visit Rosemarie George MW, president of the Circle of Wine Writers, of which I am a member. In a delightful garden behind her Hammersmith townhouse, I sampled my first-ever wine from the UK, a Nyetimber sparkler. George also had put me in contact with Julia Trustam Eve, head of marketing for Wines of Great Britain, who arranged visits to Bluebell Vineyard Estates and Bolney Wine Estate in Sussex.

Kevin Sutherland, the winemaker at Bluebell, led us in a tasting of base wines for the line of traditional method sparkling wines, as well as barrel samples for some still wines. We then sampled all of the bubblies on offer in the tasting room. I took home a bottle of the impressive 2015 Hindleap Blanc de Blancs.

We were then picked up by Helen Chesshire, Bolney's bubbly brand ambassador, and taken to the winery. After lunch, we toured the facilities, met some of the staff, and tasted a few wines including a delightful bacchus, a bottle that we took with us along with some bubbles.

Since I was researching the capacity to age of Willamette Valley chardonnay as well as how well it competes in the market against white burgundy for a piece for GuildSomm, I visited Fortnum & Mason, Harrods, and Hedonism Wines. Each only stocked a handful of Oregon chardonnays compared to a much larger selection of white burgundies.

After an all-night flight to Cape Town, we settled into a hotel on the Victoria & Albert Waterfront. Winnie Bowman, chairman of the Circle, invited us for a light dinner. Among the guests was Fiona McDonald, whom we had engaged at Winnie's recommendation to take us around to wineries for two days before the AAWE meetings. We also sipped several excellent local wines that the guests brought as well as a delicious 2005 Boplaas Brandy, which had spent seventeen years in tawny port casks.

A special treat was meeting Winnie's son, David, who had just completed his bachelor's degree in mathematics at the University of Edinburgh and would be heading off to graduate school at the University of Bonn. It was a particular pleasure to participate in postprandial patter about our passion.

Over the two days, we visited Benguela Cove, Hermanuspietersfontein, Hartenberg, Ken Forrester, Luddite, and Boschkloof with Fiona. The first two satisfied my request to sample sauvignon blancs, especially from Hermanuspietersfontein, whose winemaker had given a seminar to the Circle, while the winemaker at Ken Forrester showcased a range of classic chenin blanc. Syrah and Bordeaux varieties were featured at the other three. A standout was the 2010 Luddite Shiraz, a bottle of which was among several that followed me home.

Technically part of the City of Cape Town metropolitan municipality, Gordon's Bay is about thirty-six miles south of downtown and the site of the Krystal Beach Hotel that hosted the AAWE conference.

Though the first name of the organization is "American," the membership is international. In attendance were researchers, many from academia, from around the world. The four-and-a-half-day yearly event has settled into a schedule that begins with an evening reception, followed by two days of presentations and two of wine tasting in the area. For the first time since I've attended these conferences, I didn't prepare a talk, instead resting on my laurels after earning a Best Presentation award at the Tbilisi conference the previous year.

Following the plenary session that included welcomes and a keynote address, "An Overview of the South African Wine Industry," presented by Rico Basson, the CEO of SA Wine, separate sessions were held. Sixty-two talks were scheduled in parallel sessions on a range of topics including wine tourism; history; consumer choice, exports, markets, and prices; marketing, markets, and prices; productivity and innovation; beer, cannabis, and Prosecco; wine ratings and wine sensory analysis; vineyards and cellars; industry analyses; consumer choice; production and science; and legal issues.

Jeff Bodington has been taking a novel approach to wine scores. He contends that when a judge assigns a rating to a wine, it really is a single point in a distribution. He has come up with a way to suss out that distribution, which he discussed in a presentation with the arcane title "Almost no data – Tests of a maximum entropy estimate of the latent distribution of one wine rating," in the session entitled "Wine Ratings and Wine Sensory Analysis" that he chaired. Jeff and I have been discussing his work in light of my ongoing interest in the application of the Borda count to wine competitions and have come up with an idea on how to merge them.

Local wines were poured at the lunches and dinners and in between, so no one went thirsty waiting for the two days of winery tours. Since we hadn't included any pinotage producers on our tour before the meeting, it was nice to sample several during breaks between sessions. One standout was an older one from Kanonkop, a winery we had visited in 2013 that has become renowned for its age-worthy pinotage.

On the first day of touring, we were assigned to the Swartland Route. The first stop was Kloovenburg Wine & Olive Estate where they poured 2022 Sauvignon Blanc; 2020 Eight Feet Red, a blend of Rhône varieties; 2020 Shiraz; and 2022 Lledoner Pelut, a mutation of grenache. Next was a seated tasting at Diemersfontein where we selected wines from four "ranges" representing different styles, varieties, and price points. The tour concluded at Fairview in Paarl, which bills itself as "First and foremost, ...a working farm, housing a collection of micro-businesses," one of which is a serious producer of a vast array of varietal wines at various prices. Following a visit with one of the resident goats, the living incarnation of the "Goats Do Roam" label, we enjoyed a seated tasting followed by dinner in the Goatshed.

Stately Waterford Estate in Stellenbosch was the first stop on the second day of touring. Cellar Master Mark Le Roux hosted a comparative tasting of six cabernet sauvignons from wineries around the area. It was revelatory. The following were sampled in order of intensity: 2019 Delheim from Simonsberg, 2020 Oldenburg Rondekop Per Se from Banghoek, 2019 Stark-Condé from Jankershoek, 2018 Waterford Estate from Helderberg Mountain, 2020 Jordan Long Fuse from Stellenbosch Kloof, and 2020 Kaapzicht from Bottelary.

While each had its virtues, the one from the host winery was sufficiently pleasing that I asked if the most expensive wine they made was available to taste. Happily, one was opened for us. Considered the estate's "iconic flagship blend" and named for the owner, 2014 The Jem contains cabernet sauvignon, shiraz, cabernet franc, merlot, petit verdot, mourvedre, sangiovese, and barbera. So good it was that I added this gem to my stash despite costing over $100. AAWE Executive Director Karl Storchmann was equally impressed and took three bottles.

The second and final stop was at the Spice Route Destination in Suider Paarl. There we could taste meats, beer, chocolate, and spirits as well as wine. Jewell's Restaurant attempted to finish us off with a multicourse feast accompanied by wine.

It was back to Cape Town the next day for a last bit of sightseeing, including a cable car ride to the top of Table Mountain where we hiked around. Then we endured the two-hop trek home, alleviated with the help of complementary admissions to British Airways lounges. In tow was a full suitcase tailored for wine bottles.

Will we ever return to South Africa? It was even more beautiful than I remembered, and the wines we sampled were more impressive, with all things vinous on a clear upswing. So who knows.

PART III

Book Reviews

Voodoo Vintners: Oregon's Astonishing Biodynamic Winegrowers by Katherine Cole, Oregon State Press, 2011, 192 pp., $18.95 (paperback), ISBN 978-0-87071-605-8.

In August 2010, a fundraising event, Discover McMinnville AVA (American Viticultural Area), commenced with a seminar exploring "The Difference Between Organic, Sustainable, and Biodynamic Farming with Insight into the Geology & Terroir of the McMinnville AVA." Scott Neal of Coeur de Terre Vineyard, Robert Brittan of Brittan Vineyards, and our host, Moe Momtazi of Momtazi Vineyard and Maysara Winery, each represented one of the three approaches. The first of twelve chapters of Katherine Cole's marvelous and accessible read recounts the story of Momtazi's rapid departure from Iran with the suspense of an exciting page-turner. Moe and his family eventually settled in McMinnville, Oregon, established his eponymous vineyard, and began to farm it organically.

As Cole relates, he came under the influence of Andrew Lorand, an acolyte of Rudolf Steiner (1861–1925), the founder of biodynamics in the 1920s, and has since become one of Oregon's leading practitioners and true believers. At the seminar, Moe passed around a cow's horn he had stuffed with manure and buried underground over the winter. When exhumed in the spring, the contents, which Cole notes "looks like finely pulverized coffee grounds and smells and feels like soft, rich earth," are added to water in tiny quantities to create Preparation 500, one of nine that distinguish the biodynamic approach. While the odor was certainly inoffensive after its hibernation, Moe allowed that even after washing his hands thirty-two times after packing the horn, he could not get rid of the manure smell.

In her well-researched narrative, Cole explores biodynamics with thoroughness, wit, intelligence, sympathy, objectivity, and healthy skepticism, reflecting the wisdom obtained over nearly ten years chronicling the Oregon wine scene for *The Oregonian* and *MIX* magazine. We are given enough details about the biodynamic approach to make up our own minds about its merits, without having to endure its seminal text, the transcription of Steiner's 1924 lectures entitled *Spiritual Foundations for the Renewal of Agriculture.*

"[T]hey waiver from the general to the specific, from the tangible to the otherworldly and weird, as in 'gnomes, undines, sylphs, and fire spirits are actively involved in plant growth'" she observes. True believers, naysayers, and those in between are given full voice in her pages.

So, why is such weirdness taking hold in Oregon in the twenty-first century? Well, one reason is that biodynamics is practiced by many of the best producers in Burgundy, the wines that represent the gold standard for pinot noir and chardonnay. This clearly resonates in the New World's best response to that region.

As Cole reminds us: "For Willamette Valley winemakers, the pinnacles of achievement are the grand pinot noirs of Burgundy. And for Burgundians, the pinnacles of achievement are those grand cru vineyards that are farmed biodynamically."

Cole also argues that going biodynamic can resonate with the Oregonian personality. In 1999, Cooper Mountain Vineyards in Beaverton, Oregon, became the first to be certified by Demeter USA in the Beaver State. Since then, fourteen additional vineyards in Oregon have followed suit, with approximately an equal number that aren't certified but claim to practice biodynamical farming.

Cole concludes: "Dr. Robert Gross [founder of Cooper Mountain] may not be the most notable of the Oregon biodynamic vintners, but he was the first. In a time when 'biodynamic' was an unfamiliar term for American wine consumers, Gross made a move that showed reckless disregard for the bottom line. In his determination to stubbornly slog down an unbeaten path, he was acting the role of the quintessential Oregonian."

It is, in fact, harder to imagine a soul greener than that of an Oregonian. Many of the elements of biodynamic farming are consistent with those of other approaches under the general heading of sustainable agriculture, something many winemakers, especially in Oregon, are embracing in one manifestation or another.

"Even among the many Oregon vineyards that aren't officially certified, sustainability is both a priority and a lifestyle" proclaims the oregonwine.org website. The site goes on to cite six organizations, certifications, and programs that advocate sustainable agriculture. These are Oregon Certified Sustainable Wine®, Demeter Certified Biodynamic®, Low Input Viticulture & Enology (LIVE), Oregon Tilth Certified Organic, Salmon-Safe, and Vinea, The Winegrowers' Sustainable Trust.

The term "carbon neutral" has recently entered the vocabulary of many Oregon wineries as well. Cole reports that "[a]pproximately 30 percent of Oregon vineyards are certified by one of the state wine industry's many sustainable-winegrowing initiatives…with countless others practicing eco-friendly farming without certification paper."

The definition of sustainable not only encompasses the treatment of the soil, flora, and fauna of a site but can also include the workers and the business. When one practices sustainable agriculture, one's "eco"-system comprises much more than it had in the past since both the ecology and the economics of the farm are considered.

Demeter, the certifying authority for biodynamics founded in Europe in 1928, defines "Biodynamic farming… [as] a holistic and regenerative farming system that is focused on soil health, the integration of plants and animals, and biodiversity." No mention is made of the welfare of the workers or the viability of the farm as a business. In contrast, the LIVE program, for example, embraces these elements as well.

I particularly resonated with Cole's skepticism of the mystical aspect of biodynamic farming and appreciated her citation of credible academic authority to debunk some of its more extreme claims.

"No one argues that the 'bio' components of biodynamic agriculture…are not good, sound farming practices. It's the 'dynamic' part that loses people," she stresses. "There is some real research being conducted on biodynamic agriculture, and it's important to separate this from the random bits of pseudoscience culled from other areas of, um, academia."

There seems to be no question that, with its emphasis on regenerative farming, biodynamics is sufficient, but not necessary, for good stewardship of the land.

I was once asked by a skeptical colleague what "gourmet" means. I responded, "Care is taken." Certainly, this is the case with biodynamics. What has turned many off, however, is the manifestly antiscientific aspects, the defense of which necessarily moves into the irrational.

At the McMinnville AVA seminar in 2010, it was disorienting for me, having been trained in astronomy and physics, to listen to Moe, who despite a technical background of his own, talk about the influence on the vineyard of the positions of the stars and their use to decide when to pick. Cooper Mountain's website offers this: "We all know that the moon has an effect on the force of gravity. If the moon plays this role, it becomes logical that the solar system plays a role as well." Really?

So, how do wines from biodynamic vineyards taste? Cole's response: "Biodynamically farmed grapes make fascinating wines. They also make banal wines. The same is true of conventionally farmed grapes, organically farmed grapes, and everything in between."

I agree with her. Of those produced from biodynamically farmed grapes, I find most fascinating the wines originating in the Brick House Vineyards in the Ribbon Ridge AVA, which was Demeter-certified in 2005. They are amongst the most Burgundian, offering purity and understated complexity. I also especially enjoy the elegant wines from Maysara, the Momtazi family winery. On the other hand, I have been underwhelmed by the offerings of Cooper Mountain.

After sampling wines from dozens of producers in the Willamette Valley AVA and each of its sub-AVAs, I believe that factors other than biodynamics, such as terroir, determine the quality of the final product. But can we even tell whether a wine was made from biodynamically farmed grapes or, as Cole wonders, is "biodynamic agriculture…the key to unlocking *terroir*?"

An interesting but difficult-to-replicate experiment was conducted on July 8, 2010, and reported in the *Oregon Wine Press* the following month (http://oregonwinepress.com/article?articleTitle=holistic+tas te+test--1280776145--493--).

Twelve tasters representing a range of experience participated in a "triangular blind tasting" of two wines—one made from biodynami- cally farmed grapes and the other followed the LIVE program. All other variables including region, soils, vineyard orientation, grape variety, and winemaking techniques were the same. Each taster was presented with three flights of three glasses of wine, two of which contained the same wine.

"Although tasters disagreed upon the description, they agreed on which wine was different. …. Overall, there is no doubt that the tasting panel detected differences between Biodynamic- and LIVE- sourced wines."

Despite my strong doubts about any unique merits of biodynamic farming and my aversion to its bizarre antirational practices or, as Cole labels them, "extraneous spiritual baggage," I admit, unblush- ingly, that I loved this book, which gave me detailed information on both the theory and actual application of this controversial approach. Having recently settled in McMinnville and immersed myself in its wondrous world of wine, I have come to know personally many of the characters mentioned and have even met and briefly corresponded with Cole. Undeniably, this added to my enjoyment.

The vast majority of the winemakers mentioned produce pinot noirs that invariably transport me to new levels of pleasure, regard- less of how the grapes were farmed. To gain the fullest appreciation of biodynamics for yourself, I recommend a superb wine and food- for-thought pairing—a glass of Brick House or Maysara pinot noir and Cole's monograph.

Since this review first appeared in 2012, other wineries in Oregon have adopted biodynamic practices, though many have not pur- sued Demeter certification. Another movement that has gained a handful of adherents from the wine world is B Corporation certification.

From its website: "Certified B Corporations are businesses that meet the highest standards of verified social and environmental performance, public transparency, and legal accountability to balance profit and purpose. B Corps are accelerating a global culture shift to redefine success in business and build a more inclusive and sustainable economy." (https://www.cultivatingcapital. com/b-corporation/)

A few wineries have been experimenting with regenerative farming, which has its own certification program: "Regenerative Organic Certified® is a revolutionary new certification for food, textiles, and personal care ingredients. Regenerative Organic Certified® farms and products meet the highest standards in the world for soil health, animal welfare, and farmworker fairness." (https://regenorganic.org/)

If anything, the Oregon wine industry is increasingly emphasizing careful stewardship of the land and reduction of its carbon footprint as the effects of climate change have become more manifest. The resulting juice seems to be improving as well.

What follows comes with a story. The Journal of Wine Economics (JWE), *which has liked (yea verily, loved) two previous book reviews, offered to get me a copy of any wine book I would like to review for them. The first one sent to me was* How to Love Wine *by Eric Asimov. I prepared a review and here was the response from the editor, Karl Storchmann:*

"I ran you're [A]simov review with my co-editors. Orley [Ashenfelter] and I love it.

But two others think it should be toned down and not too obviously smash [Asimov].

Can you do that?"

And so with my wife, who is much nicer than I am, I did and resubmitted it. Here was the next response from the editor:

"I am sorry to let you know that there is a strong opposition to your book review on our board that I have to turn it down. Some feel that it is 'aggressive and almost personal.'

I am really sorry about this."

So, I immediately sent the original version to the Journal of Wine Research (JWR). *Here was their response:*

"Many thanks for your book review, which we are pleased to accept in its current form and which will now be forwarded to the publisher for copy editing and typesetting. The book review editor's comments are included at the bottom of this letter."

"Book Review Editor Comments to the Author: Fully entertaining and engaging, as well as thorough and well written – we will be delighted to have this piece in the Journal, and thank you for the submission."

I guess one journal's meat is another's bane.

I've adopted a modified version of advice my parents gave me: If you can't say anything nice, at least write it down. Enjoy!

how to love wine: a memoir and manifesto by Eric Asimov, New York, William Morrow, 2012, 278 pp., $24.99 (hardback), ISBN: 9780061802522.

There is a high probability that if you are reading this, you already love wine. So is there any reason for you to read a book that purports to teach you how to do something you already do? And if you don't already love wine, will this book lead you to a path that will arouse such affection? My answer to both questions is a tepid "maybe, but probably not."

Let's deal with the second question first. The author, Eric Asimov, chief (and sole) wine critic of the *New York Times*, begins his outreach to the unloving with a bit of therapy. He observes that "…what's missing in many people's experience of wine is a simple sense of ease. Instead, choosing a wine becomes an exercise in anxiety…" (p. 3).

While not nearly as damaging to the societal fabric as, say, math anxiety, "wine anxiety" is a genuine affliction that one should overcome if one is to develop warm feelings toward *fermentatum uvae suci*. In his role, Asimov is frequently asked for resource recommendations that will teach folks what they need to know by cutting through the layers of well-encrusted bullshit that has occluded wine appreciation for decades.

Hedges Asimov: "This book will not demystify wine for anyone… What I really hope to do is to clear up the murky, intimidating business of enjoying wine…" (p. 7).

But before we proceed, we must first attain the proper mindset. "Personally, I feel that wine cannot be understood without a sense of uncertainty, of modesty and humility and respect," (p. 33) stresses Asimov. Could this be the reason for the e e cummings-esque lowercase title on the first edition of the book and an entire chapter dedicated to "The Importance of Being Humble"?

What, then, are the key elements of Asimov's guide to the vinously perplexed?

"The single most important thing one can do…is to make friends with a smart salesperson at a good wine shop…" (p. 9).

Okay, I'll buy that. To this day, I do solicit the recommendations of merchants and sommeliers who have earned my trust by delivering numerous delights to my table.

In a chapter entitled "Seeking Higher Learning," Asimov relates his experience taking a class at a wine school in New York.

"I left the class puzzled by what exactly I was learning to do there…the whole process seemed sterile compared with the way I actually drank wine." (p. 144). Instead, he recommends that you "[h]ome-school yourself" (p. 209). Echoing what he acknowledges is a "hackneyed suggestion" (p. 209), Asimov writes: "The best way to learn about wine is to drink a lot of it…" (p. 209).

He suggests asking your friendly neighborhood salesman to assemble a case of red and white wines costing about $250. "My feeling has long been that $15 to $20 a bottle is the sweet spot for great wine values. Below $10, it's easy to find wines that are sound, but a struggle to find bottles that inspire." (p. 211).

And that, folks, is the distilled wisdom of this tasting Torah, easily absorbed while standing on one foot. Now go forth, open some bottles, and start sipping.

As someone who has been rapturously volatizing vinous esters since 1968, I was most interested in Asimov's accounts of visits to vignerons and negotiants, many of whom are little known. While lacking the emotion, drama, and human interest of Sergio Esposito's *Passion on the Vine*, descriptions of encounters with several European and California producers and their products made for some of the better reading.

I share the author's admiration for the wines of Joh. Jos. Prüm. Reminiscing about a 2002 Domaine du Jaugaret from a tiny vineyard in Bordeaux, Asimov declares: "It was perhaps the most direct expression of St.-Julien terroir that I've ever had…[t]hat wine haunted me…"

(p. 232). Those who already love wine might find some pleasure in reading the all-too-brief reflections of someone who does get around.

Asimov prefers to regard wine as "an expression of culture, … [which] differs significantly from those fields we consider as the arts…" (p. 222). While I agree that "[t]he art of producing wine…is much more of an interpretative performance" (p. 222), I take issue with "Wine itself is not art, at least not in the same way as music or paintings." (p. 222).

I have had the same emotional and aesthetic response to a remarkable bottle as I have had to a great symphony or a well-executed canvas. Despite making the above unambiguous assertion, Asimov appears to agree with my view when he waffles: "Yet wine does have the capacity to move us…wine can display great beauty…" (p. 225).

Since in many instances, art is an expression of culture, there really is no need to categorize wine as one or the other.

A chapter, "The Tyranny of the Tasting Note," so effectively trashes the pervasive practice of ejaculating nouns, adjectives, and adverbs in an ultimately futile attempt to capture the experience of enjoying a glass of wine that it may have obviated the need for me to complete an article I began on the same subject.

Asimov is ruthless in his criticism: "At best, tasting notes are a waste of time. At worst, they are pernicious." (p. 85). Right on, I say!

On the other hand, in "Drinking by Numbers," he is somewhat less critical of the 100-point scoring system: "…as clearly as I see the unintentional damage that tasting notes inflict, I am a little more ambivalent about the issue of scoring wines… Wine scores can interfere with consumers developing their own standards and preferences…[but] numbers cut through the confusion caused by thousands of bottles and the mysterious words used to describe them…" (p.164).

I object to this for two reasons. First, it exemplifies one of my objections to his manner of exposition that permeates the book and that I have already hinted at—namely, an unwillingness to take an unpopular stand or assert a firm opinion without qualification.

Second, I steadfastly believe that point scores for wine are the devil's spawn. It turns out that Asimov makes my case in the balance of the chapter, concluding that "scores are a poor substitute for wisdom..." (p. 175). Which begs the question: Why, then, the ambivalence?

Interspersed with his observations about the state of the wine world and his recommendations for finding your place in it, Asimov chronicles his own journey to wine nirvana. One has no control over the circumstances of one's birth including place, family situation, and such. Frankly, Asimov's is fairly mundane and even somewhat bourgeois. There was no flight to a better life with its concomitant adjustments to language and culture and struggles to make a living as in Esposito's memoir. Recounts of growing up in "Roslyn Heights, a pleasant, shady Long Island suburb" (p. 56) are mostly a yawn flecked with occasional examples of too much information.

For a dozen years before being named as Frank Prial's replacement in 2004, Asimov wrote about inexpensive restaurants in a column in the *Times* entitled "$25 and Under." But compared to the escapades as a camouflaged critic chronicled in Ruth Reichl's *Garlic and Sapphires*, Asimov's adventures are given the shortest shrift, only about four pages. Ungraciously, he even takes a backhanded slap at his former colleague at the Old Gray Lady: "I thought of a disguise as the last resort of the exhibitionist..." (p. 187). He may call this modesty, but Reichl's tales are orders of magnitude more entertaining.

Organizationally and structurally, the book is unsatisfactory. "I feel that the various parts of this book may not always dovetail neatly..." (p. 13) apologizes Asimov. True enough.

The copyright page acknowledges that "several passages were adapted from work that appeared previously in the *New York Times.*" Unfortunately, the texts are not always comfortably harmonized, and there is much unnecessary repetition of ideas with no additional insight, reinforcing a nagging feeling I had that there really wasn't enough good or original material to fill even this relatively small volume.

By writing both a memoir and a manifesto, Asimov attempted to sing "My Way" in two voices. But because there really isn't that much new or interesting to convey, neither carries the gravitas of Paul Anka or Frank Sinatra. So, if you are an expert or a wannabe, you are best advised to allocate the price of this book to a bottle of something you have never tried before.

Best White Wine on Earth: The Riesling Story by Stuart Pigott, New York, Stewart, Tabori & Chang, 2014, 208 pp., ISBN: 978-1617691102 (hardback), $24.95.

Sometimes I imagine I can still taste it. The deep dark brownish, apricot-hued 1959 Steinberger Trockenbeerenauslese, with the rich intensely fruity bouquet burst onto my palate with remarkable flavors like biting into a perfectly ripe honeyed apricot. It lingered with a depth of finish I had never before experienced, so my contemporaneous notes recount.

This nectar concluded a tasting of aged clarets I hosted on May 22, 1977, to celebrate the completion of my doctorate. My gourmet group's practice was to serve an Auslese when one was awarded a bachelor's degree and a Beerenauslese after earning a master's. During the 1970s, we had tasted several other bottlings from the Steinberg vineyard near Hattenheim, my notes on which are among the most effusive. This isn't surprising since the wines from that vineyard were called "the kings of the Rheingau" by Frank Schoonmaker in his classic *The Wines of Germany*.

My, how the world—or "Planet Riesling," as Stuart Pigott prefers to call it—has changed. In his essential and exuberant paean to the best white wine on Earth, he chronicles the past quarter century's remarkable evolution of "my favorite grape," as he frequently refers to riesling, without any color distinction.

This hymn of praise, however, isn't written in iambic pentameter. Instead, it reflects his penchant for Gonzo journalism "that is written without claims of objectivity, often including the reporter as part of the story via a first-person narrative" (http://en.wikipedia.org/wiki/Gonzo_journalism). As a homage to Pigott and his influences, Hunter S. Thompson and Tom Wolfe, I have adopted the same style for this critique. After all, what is good enough for the author should be good enough for the reviewer.

Pigott is all over the narrative and illustrations. He coins his own terminology. The global network of wine professionals around new rieslings is "Planet Riesling" (p. 13). He refers to "Blade Runner steeliness" (p. 17), then uses the term to designate a separate category of rieslings distinct from the more common dry, medium dry, medium sweet, and sweet. Inexplicably inappropriate selfies appear in a couple of places (p. 13 and p. 189). Tales of his personal encounters with many of the most noteworthy producers around the globe comprise most of the text.

It is on this trip around the new world of riesling that the book is most valuable. After introductory chapters that briefly present the history of the grape and describe the various styles of wines made from it, the tour begins in the wine lakes of the Northeast. I have limited experience with the products of New York's Finger Lakes—or FLX as Pigott prefers to call it—an area I visited only once. So, it was nice to learn that Dr. Frank and Hermann Wiemer, while both still significant players, weren't the only ones in town. The recent development of the wine industries in Ontario and Michigan made for good reading.

We next head to the West Coast. California, of course, plays prominently, but Oregon merits an insightful discussion. While pinot noir reigns supreme, especially in my home region, the Willamette Valley, riesling has almost as long a history with the first vines planted over forty years ago. Some of my favorite producers—Chehalem Winery and Brooks Winery—are among those singled out. I have also had remarkable examples made from grapes grown in the Maresh Vineyard in the Dundee Hills American Viticultural Area (AVA) and Hyland Vineyard in the McMinnville AVA, both amongst the oldest in the area.

While in the West, Washington State's Chateau Ste. Michelle— "the world's biggest producer of riesling and the most consistent of those producing riesling on this grand scale," (p. 76)—is an obvious stop. Canada's Okanogan Valley in British Columbia is the home to a handful of producers whose individual approaches led Pigott to observe that "Okanogan Riesling really can't be reduced to a simple formula" (p. 82).

I have marveled at Grosset Polish Hill Rieslings from the Clare Valley in South Australia the couple of times I have had them. This winery is one of several that Pigott takes us to in OZ, as he likes to call it. So busy was I sampling just about every other variety that I don't recall tasting any rieslings during my extensive tour of New Zealand wineries in 2004. As I learned in the book, the distinguished efforts of a handful of winemakers are eclipsed by the attention lavished on sauvignon blanc.

Austria merits its own chapter with extensive discussions of the areas around the Danube. In the land better known for grüner veltliner, riesling "has gone from being a specialty to one of the most important white grapes for quality wine in Austria during the last generation" (p. 116). The wines tend to be bigger, with higher alcohol than many from Germany.

We finally arrive in the land of the Rhine and its tributaries, the home of the riesling grape. The chapter begins with a statement that made me aware of my age: "Riesling, and particularly German Riesling, has long suffered from the image of being old fashion [sic]…" (p. 129) and invariably sweet. During the same period, I savored the Steinbergers, I consumed many crisp Kabinetts, intriguing Spätlesen, glorious Auslesen, and, on the rarest occasions, luscious Beerenauslesen from some of the finest producers, several of whom I visited on a trip to Germany in 1978. Along with burgundies and clarets, these superb German rieslings set the hook, as it were, and committed me to a lifelong love of wine. They may be viewed as old-fashioned, but to me they were revelations.

After a quick history that explains some of the basis for the reputation, Pigott turns his attention to what has been happening lately, not only among many of the outstanding winemakers I came to appreciate but also among the new breed known as Jungerwinzer—"a word that now doesn't only mean young winemakers but also implies talent, creativity, and coolness…" (p. 130). Alsatian rieslings, in my experience amongst the driest and most versatile at the table, are covered in the same chapter as the Germans since the area is on the French Rhine.

The penultimate chapter covers Riesling Lone Rangers, countries where the grape is grown but has not gained prominence. Efforts in Italy, Eastern Europe, South America, and South Africa are lightly touched upon.

"The Stuart Pigott Riesling Global 100" concludes the exposition. While Steinberger is nowhere to be found, though there is an oblique reference to it in a discussion of Kloster Eberbach, the monastery where the wines are made (p. 134), several of my old and new friends make his list. For those he highlights that aren't familiar to me, having this calibration suggests new things to try.

Wine economists should be amused by the 1949 wine list from Houston's Shamrock Hotel, listing higher prices for Rhine and Moselle wines than for most clarets. Insets throughout the text offer "Crazy Riesling Stats" galore.

While this volume is the best contemporary account of the state of riesling I know of, it is not without some distractions. There are some instances of poor editing including typos, spelling errors, and run-on sentences. I wish the book had a glossary and an expanded indexing to facilitate searching for terms. Instead of maps of prominent riesling growing areas and pictures of labels beyond those on p. 148, there is an inordinate number of photos of winemakers and others in the trade. While most of the pictures of vineyards are a welcome break from the densely worded, two-column pages, I find some, like the one spanning pages 182 and 183, uninspiring.

These defects aside, I know of no other writer who is more qualified to extol the virtues of this most important variety. Pigott summarizes his case for his favorite grape: "…the entire point of Riesling is the wines' diversity, and as long as they are well made, this diversity is enriching…" (p. 162). The book provides ample evidence and thus merits consideration by oenophiles of every degree of experience.

Whereas Pigott makes his preference crystal clear, I am still occasionally conflicted as to what I consider the greatest grape—

riesling or pinot noir. A brief conversation at the 2014 Passport to Pinot with Wynne Peterson-Nedry of Chehalem, a noted producer of marvelous wines from both, may eventually sway me in Pigott's direction. If we use the desert island test, she pointed out, riesling would be more appropriate to accompany what we are likely to eat. Hmmm. Compelling, but I think I'll continue to study the question.

And I still am.

Brooks continues to be the gold standard for Willamette Valley rieslings bottling a range from dry to sweet and sparkling each year. However, following the sale of Chehalem in 2017, Wynne, now the proprietor of Ridgecrest along with her legendary father Harry Peterson-Nedry, is giving them a run for their money with her offerings under the RR and Ridgecrest labels. Her 2017 RR Riesling Estate Reserve is fabulous and aging beautifully. I have also had wonderful rieslings from Hyland Vineyard first planted in the McMinnville AVA in 1971 at the dawn of the Northern Oregon wine industry.

I visited the Finger Lakes Region a second time in 2018 and was impressed with rieslings from Ravines, a winery run by Dr. Konstantine Frank alum Morten Hallgren and his wife Lisa. Dr. Frank continues to produce lovely wines. I also sampled a few rieslings during a visit to wineries in Northern Michigan in 2020 and found them pleasant but simple and a bit thin.

I'm still not drinking enough German riesling.

Winemakers of the Willamette Valley: Pioneering Vintners from Oregon's Wine Country by Vivian Perry & John Vincent, Charleston, South Carolina, American Palate, 2013, 160 pp., ISBN: 978-1609496760 (paperback), $19.99.

Oregon Wine Pioneers by Cila Warncke, Portland, Oregon, Vine Lives Publishing, 2015, 234 pp., ISBN: 978-1943090761 (paperback), $19.99.

A dmittedly, it was hard to write a dispassionate review of books that so lovingly describe the region in which I live and so admiringly profile many of my acquaintances in the Oregon wine industry. Therefore I used as measures of merit how well each echoed my impressions of this most beautiful area and its people, and whether each accomplished its objectives.

Perry's and Vincent's *Winemakers of the Willamette Valley* (WWV) "is meant to showcase the stories of a handful of Oregon's many Willamette Valley winemakers…" (WWV, p. 11). A foreword by Chehalem founder Harry Peterson-Nedry sets the personal tone that pervades those stories.

Next, in a mere eight pages of text, the first chapter, "History of the Willamette Valley Wine Region," covers the climate, soil, grape selection, craftsmanship, industry structure, and early success in sufficient detail to provide valuable context.

The authors then share intimate interviews with eighteen vintners and vignerons. Within each chapter named for one or two winemakers are brief descriptions of the wineries that each is affiliated with. These include year founded, ownership, varieties, tasting room location, hours, and contacts.

Sustainability features, a point of pride in the Oregon wine industry, are also listed. The epilogue memorializes the late Willamette

Valley Vineyards winemaker Forrest Glenn Klaffke. Wine Tasting Routes and a list of wineries by town provided by the Willamette Valley Wineries Association comprise the appendix.

In contrast, Warncke's *"Oregon Wine Pioneers* (ORWP) aims to tell a good story and inspire you to take part in that story. We hope you take it along when you head out to visit…" (ORWP, p. 9). Although the fifteen chapters are named for wineries, they contain much biographical information about the principals. Each concludes with tasting notes of wines made at the facility and a lined page for the reader's own comments. Six Trail Guides for Portland, Forest Grove, Newberg, McMinnville, Salem, and Southern Oregon follow. These give driving directions, contact information for the featured wineries, and restaurant recommendations.

"By definition, there can only be one group of pioneers" (ORWP, p. 29), Warncke tells us. From a strict point of view, then, only the first wave of a dozen producers who began coming to Oregon about a half-century ago should be regarded as pioneers.

Wisely, though, the two books de facto adopt a broader perspective. Warncke presents vignettes about winemakers who became part of the Oregon wine industry well after the 1960s and 1970s.

For example, she interviews Earl Jones, who pioneered high-quality tempranillo in the United States at Abacela in the 1990s. And WWV, which includes "Pioneering Vintners" in its subtitle, profiles folks like Steve Doerner, winemaker at Cristom Vineyards since 1992, who "is a thirty-five-year practitioner of whole-cluster, native yeast fermentations" (WWV, p. 51).

Both volumes cover Adelsheim, Elk Cove, A to Z/REX HILL, and Ponzi Vineyards, with some overlap but enough differences to make each worth reading. While winemaker David Paige is the focus of WWV's chapter on Adelsheim (WWV, Chapter 6), founder David Adelsheim is highlighted in ORWP (ORWP, pp. 22-33).

Both concentrate on Elk Cove's second-generation winemaker Adam Campbell, but ORWP also introduces his sister, Anna Campbell (WWV, Chapter 8; ORWP, pp. 34–45). WWV (WWV, Chapter

12) features Anna Matzinger and Michael Davies—the latter, the executive winemaker at A to Z/REX HILL—whereas ORWP (ORWP, pp. 46-59) takes us on a tour of that winery with the direct sales manager that includes a cameo appearance by cofounder Deb Hatcher, but no mention of Davies.

The greatest degree of overlap is in the chapters on Ponzi Vineyards (WWV, Chapter 7; ORWP, pp. 60–71) wherein Luisa Ponzi, who took over as winemaker from her father, Dick, in 1993, is the center of attention.

WWV concentrates on producers who get their grapes primarily from the six American Viticultural Areas (AVAs) that partially overlap the northwest portion of the Willamette Valley AVA. ORWP ventures further south with profiles of Illahe Vineyards in Dallas, Left Coast Cellars in Rickreall, and Abacela in the Umpqua Valley near Roseburg.

The trail guides in ORWP are much more valuable resources for the prospective tourist than what is given in WWV. In particular, I can vouch for many of the restaurants included. While the tasting room information in each chapter of WWV might be useful, it should be confirmed as things do change. The appendix, however, seems like an afterthought.

While none of the authors are established wine writers, all have published extensively, so both books read very well and very quickly. Journalistic WWV relies more on quotations and less on descriptions of the land, the processes, and the writers' personal reactions.

The writing in ORWP struck me as more literary, impressionistic, and passionate. We share moments of realization with Warncke:

"Voila. The missing piece. The link. The glue. I should have guessed. The clue is in the name: A to Z. You can say anything with 26 letters and this is a winery dedicated to expression. Climate, soil, elevation, varietals, and water, are the winemaker's alphabet…" (ORWP, p. 56).

I also enjoyed the clever analogies Warncke draws. In describing the career path of Tom Symonette of Whistling Dog Cellars, she writes: "…a picture emerges of a man whose life—like the vines he tends with such intense affection—had three buds. Two of which, removed, left one strong shoot..." (ORWP, p. 100).

WWV edges out self-published ORWP for production value with sharper photographs and affectionate sketches of the winemakers by Sarah Schlesinger. Still, the latter skillfully weaves uncaptioned snapshots into the text from which they derive their significance. Both successfully give the reader a sense of what it is like to visit a winery in the Beaver State.

There are some minor quibbles. I found the tasting notes in ORWP rather useless and even a bit bizarre—petrol notes in pinot (ORWP, pp. 20, 58, 96, 108)?! Also, many of the wines mentioned are likely no longer available. Inadvertently, no doubt, punk artist Don Letts is referred to as an Oregon wine legend (ORWP, p. 86), dislodging "Papa Pinot," David Lett.

The number of vineyards and wineries in Oregon is woefully underreported as 400+ (ORWP, p. 9). The 2014 Oregon Vineyard and Winery Census Report published by the Southern Oregon University Research Center in August 2015 reports an increase of 8 percent to 1,027 vineyard operations and from 605 to 676 bonded wineries. There is some unnecessary repetition in WWV, for example, regarding The Carlton Winemakers Studio location, fee, and contact information (WWV, pp. 83 and 85).

Books of this sort do have a limited shelf life since they report on a fluid industry. Much has changed even in the short time since WWV was released. Scott Wright sold his interest in Scott Paul Wines and Kelley Fox (WWV, Chapter 11) no longer makes wine for that label. Anthony King (WWV, Chapter 13) is now General Manager of The Carlton Winemakers Studio (WWV, Chapter 9). Don Crank III (WWV, Chapter 16) left Willamette Valley Vineyards and is now at REX HILL (ORWP, pp. 46-59). Eric Hamacher (WWV, Chapter 9) was just named winemaker at Ghost Hill Cellars (ORWP, pp. 84-97).

Before I became a full-time resident of Oregon, I spent part of the year in Virginia and would invariably miss McMinnville. I would devour each issue of the *Oregon Wine Press* when it arrived so that I could be transported back to where I wanted to be. As I read these two adoring accounts of an industry of which I am now a part, I was reminded of how lucky I am to be here and to experience daily this extraordinary place and its people.

For those less fortunate, reading both Perry and Vincent, and Warncke can give a satisfying vicarious experience. The two accounts dovetail nicely with the resulting binocular view more complete than any one of them would provide. For the price of a good bottle of Oregon pinot noir, these two slim attractive volumes will make you want to visit, if the wine hasn't already convinced you to do so.

Now, with over a decade since the publication of WWV and nearly a decade for ORWP, even more has changed, especially with the impact of the pandemic. Nevertheless, these books will retain value as snapshots of a small but increasingly important wine region during an earlier period of growth and development. They might even trigger some pleasant memories in readers in the future.

Riesling Rediscovered Bold, Bright, and Dry, by John Winthrop Haeger, Oakland, University of California Press, 2016, 384 pp., $39.95 (hardback), ISBN: 978-0520275454.

In 2014, Stuart Pigott's *Best White Wine on Earth: The Riesling Story* (Pigott, 2014) took us on an exuberant, if somewhat haphazard, global tour of what he called "Planet Riesling." My review (Hulkower, 2014) concluded: "While this volume is the best contemporary account of the state of Riesling I know of, it is not without some distractions." Because Pigott is a journalist and not a scholar, objectivity and thoroughness were not his main goals. Those, including me, who like a little rigor with their riesling, will be more than satisfied and, perhaps, overwhelmed by Haeger's latest tome.

Haeger is billed as "a sinologist, historian, and academic administrator who has written about wine since 1985." He authored two books on pinot noir, further evidencing his impeccable taste in wine. While acknowledging Pigott's and others' contributions, he delineates three goals for his most recent effort: distinguish riesling from other white varieties, reexamine the history of that grape, and provide a detailed discussion of "what must happen, both in the vineyard and in the cellar, to produce very good Riesling with little or no perceptible sugar..." (p. 3). He achieves all three.

In contrast to Pigott, who covers the spectrum of Riesling from dry through sweet in both hemispheres, Haeger concentrates on dry riesling produced above the equator. Part I comprises eight chapters that make up a little over a third of the book. Topics include the definition of dry; balance in riesling; "A History of Riesling, Reviewed and Amended;" styles; the making of dry riesling; clones; and riesling habitats in Western Europe and North America.

Part II presents fifty-six European and thirty-three North American sites that are producing noteworthy examples of dry riesling.

Individual producers who vinify riesling from these sites are also introduced. A collection of maps is a welcome feature that I regularly turned to when reading through this part.

Part I is the more reader-friendly, authoritatively presenting an overview of key topics but saving detailed discussions of specific growers and producers for the second part. Riesling is a variety "grown on every wine-producing continent…" (p. 9). With the possible exception of chenin blanc, "Riesling's many styles differ primarily in their levels of sweetness… [t]he differences in residual sweetness… [span] a mind-boggling two orders of magnitude." (p. 11)

Compared to other whites, riesling's aromatics require using the largest portion of the lexicon of wine descriptors ranging from cool climate and tropical fruits, flowers, spices, condiments, stone, and "the controversial…petrochemical note…variously described as petrol, diesel or kerosene." (p. 24) The source of this oddity is 1,16-trimethy-1,2-dihydronaphthalene or TDN. Haeger's discussion of this chemical in Part I and references to it in Part II gave me the most complete explanation I have read.

The third chapter, "A History of Riesling, Reviewed and Amended," addresses Haeger's second goal. The first mention of the grape was in 1435, though the reliability of this date has been questioned. One of its parents, gouais blanc, was identified in 1998 via DNA fingerprinting, but the other remains elusive. This, of course, causes Haeger's historian sensibilities to kick in, resulting in an eight-page discussion of documentary evidence to draw the "Big Picture." Not surprisingly, more questions—such as why riesling was selected over other varieties "for planting in Rüsselsheim and Trier in the fifteenth century" (p.41)—remain unanswered. Nevertheless, one can marvel at this illustration of how a wine historian's mind works.

Mercifully, in the first part, Haeger breaks up the narrative contained in dense, two-column pages with over a dozen boxes detailing special topics. The first addresses "How is 'dry' defined?" The description of dry riesling is complicated by the fact that the perception of sweetness is influenced not only by the amount of sugar remaining after fermentation but also by acidity which counterbalances it and alcohol which reinforces its perception.

"Within certain thresholds, more acid makes the same amount of sugar taste less sweet..." (p. 13). For the most part, Haeger adopts the European Commission definition: a wine is dry if it has no more than 4 grams per liter (g/l) of sugar (0.4%) or no more than 9 g/l of sugar (0.9%) if it does not exceed the acidity by more than 2 g/l.

I particularly enjoyed Box 6A, "Tasting Clones in Oregon." It is a wine-wonk's tour through six clones of riesling planted by Harry Peterson-Nedry of Chehalem in the Wind Ridge Vineyard in the Ribbon Ridge American Viticultural Area. Haeger includes tasting notes from December 2014 of samples of each clone harvested and vinified separately in 2013. Only one of the clones "made a 'complete' wine on its own." (p. 76).

Among the other boxes is one that wrestles with the notion of balance and several that go into the arcana of vineyard names and appellations. One is dedicated to Chateau Ste. Michelle in Washington State, the largest producer of riesling on Earth.

The single criterion for the inclusion of particular vineyards and wineries in Part II is the existence of "parameters of [the] site that are expressed in finished wine and especially those that bear on success with Riesling made dry." (p. 3). Here we learn about venerable vineyards and wineries as well as ones barely established. This part covers five regions: the Rhine Basin including the Alsace and Germany regions; the Danube: Lower Austria; the Adige Basin: Alto Adige; Eastern North America, and Western North America, including Okanogan and Similkameen; Washington and Oregon; and California's Coastal Valleys.

In each section, a site's location, age, orientation, size, soil, plantings, history, and ownership are given, followed by information about one or more producers who make dry rieslings from grapes grown there. For example, in the Alsace subsection of the Rhine Basin, we learn about the Rosacker and Clos Ste-Hune sites. Three producers using fruit from these sites—Vins D'Alsace Mader, Domaine

Mittnacht Frères, and Domaine F. E. Trimbach—are then profiled. Usually, at the end of a site's write-up, Haeger includes tasting notes of several vintages made from it by a noted producer.

Because the writing style in Part II shifts from technically ponderous to thirst-inducing across contiguous paragraphs, the book is challenging to read in anything other than small doses. Consider the juxtaposition of "…he…replanted…with tighter spacing (1.7 by 0.6 meters), using *massale* selections of scion material grafted to 3309 rootstock" (p. 198) with "The 2008 was bright, very elegant, and lovely with strong minerality, high-toned citrus, and terrific freshness of flavor…" (p. 199).

Haeger's obsession with admittedly admirable academic precision can lead to some amusing if mind-numbing descriptions: "Thus three selections from Rauenthal vineyards…are known as Rauenthal 69, 95, and 98 or as Rauenthal 69 Gm, 95 Gm, and 98 Gm…" (p. 73).

Such instances abound. For those of us who are not viticulturists, they exemplify the term "boredons" (The Onion, 2008), subminutiae devoid of useful information, and broaden the appropriateness of the word "dry" in the title.

Another concern is the time sensitivity of the information included, especially about relatively new producers who haven't yet fully established themselves. Certainly, in the New World, the wine industry has very few examples of family businesses going back more than a generation or two. More typical are producers who rely on talent who are more mobile and may stay only a few years before moving on.

Two sites in Oregon profiled in Part II provide contrasting examples. Elk Cove Estate Vineyard, one of the original pioneers in the Northern Willamette Valley, is now under the guidance of the second generation, which is completely involved in growing and production. Lemelson Vineyards, which debuted in 1999, has brought in a new winemaking staff since the departure of Anthony King, who is mentioned in the book.

Nevertheless, despite these issues, this systematic and meticulous volume, in so many ways an impressive work of scholarship, belongs on the shelves of anyone with either a professional or consumer interest in wine. Viticulturists and winemakers who want to compare notes with other riesling growers and producers can start here. Wine writers will find it an incomparable reference. Wine tourists will learn of new as well as well-established vineyards and wineries specializing in dry riesling to visit in Europe and North America. The fifteen-page, two-column index facilitates navigating the complex text.

Despite the unsurpassed depth of his book, Haeger acknowledges it isn't complete: "With regret, I have confined coverage to the Northern Hemisphere…This decision, compelled by considerations of time, distance, and expense, has made it possible to finish this book within five years of its beginning but, alas, not to finish the story…" (p. 3).

I hope that the response to this extraordinary work affords Haeger the resources to complete his tale. For those who can't wait, Pigott's book can fill the gap nicely in the interim.

References

Hulkower, N. D. (2014). Review of Best White Wine on Earth: The Riesling Story. *Journal of Wine Economics,* 9(3), 358-361. doi: 10.1017/jwe.2014.35.

Pigott, S. (2014). *Best White Wine on Earth: The Riesling Story.* New York: Stewart, Tabori & Chang.

The Onion. (2008). Researchers Discover Details Smaller Than Minutiae. http://www.theonion.com/article/researchers-discover-details-smaller-than-minutiae-6186.

*Haeger has not published anything about
Southern Hemisphere riesling.*

I Taste Red: The Science of Tasting Wine, by Jamie Goode, Oakland, University of California Press, 2016, 224 pp., $29.95 (hardback), ISBN: 978-0520292246.

One can view wine simply as an adult beverage that pairs particularly well with food and provides pleasure to the consumer. For some folks, this description is all that is required. If, on the other hand, one wishes to delve deeply into what happens when one takes a sip, swirls, and swallows, there is a growing body of scientific literature available, albeit not necessarily suitable for the nonspecialist.

Recently, two books have been published that are intended to bridge the gap between the academic literature and the interested but nontechnical wine lover: *Neuroenology: How the Brain Creates the Taste of Wine* by Yale neuroscientist, Gordon Shepherd, (Shepherd, 2017) and Jamie Goode's *I Taste Red: The Science of Tasting Wine*. While neither of these falls under the category of light reading, the Goode book is more successful on several levels.

With credentials in a hard science, plant biology, a long list of awards for his wine writing, and experience as a wine judge, Dr. Jamie Goode is particularly well qualified to take on the challenge of explaining what goes on, not only in the nose and mouth but in the brain of a wine taster. More compellingly, he makes the case as to how this information can make one a more attentive and, therefore, a more appreciative taster.

In the introduction, Goode writes: "I do not dispute that some people are gifted tasters...but I am not convinced that there is always a single true or valid interpretation of any particular wine." (p. 5)

Different biological equipment, experiences, and aesthetic sensibilities can be expected to yield different reactions. On the other hand, while personal, wine tasting is not completely subjective. It is a special example of the multimodal perception of flavor that, to some extent, can be trained. Thus, "I am going to be arguing that it is time the wine trade adapted and modernized its model for wine tasting." (p. 5)

Goode lays the foundation for his case in the first nine chapters and outlines his recommendations in the tenth and final. Expanding on the book's title, the first chapter, "What Does Red Taste Of?" describes synesthesia, the phenomenon where the stimulation of one sense results in the stimulation of another as well. It is an extreme and aberrant example of how our senses don't act separately, a fact that is at the core of multimodal perception. Synesthesia is studied in part to understand the "binding problem," how the representations in different parts of the brain of a single sensory experience blend into a single perception.

Chapter 2 focuses on the two chemical senses, smell and taste. Smell comprises two sensations, orthonasal olfaction in the nose and retronasal olfaction from the back of the mouth. The olfactory epithelium, which contains the olfactory receptor neurons, detects smell when breathing in via orthonasal olfaction and breathing out via retronasal olfaction. These signals are sent to the olfactory bulb in the brain, which then produces smell. How all this happens is still unknown. Decomposing the phenomenon into receptor space and perception space must include taking account of interactions. Much experimental work is underway to sniff out the details.

Goode debunks the misconception that taste receptors are localized on the tongue in different areas, attributing it to a misunderstanding by a translator of a German study. Instead, taste receptors are spread evenly on the tongue, which is also touch-sensitive. Smell, especially retronasal olfaction, taste, and touch—along with the other two senses, hearing and sight—join in the multimodal perception of flavor.

This realization, as it applies to wine, is further explored in Chapter 3. Goode states, "[T]here is a distinct sense called flavor, which is the result of the brain combining information from taste, smell, touch, vision and even hearing…" (p. 55). After much discussion about brain models and brain processing, he addresses how flavor is created. "Flavor seems to be complicated and multimodal" (p. 65), he suggests, before summarizing studies done using functional magnetic resonance imaging (fMRI) to trace blood flow in the brain while sipping wine.

Despite being conducted in a distracting and unnatural environment, the experiments are yielding new data that are forming the basis for our understanding of how the brain translates signals from the senses into smells and flavors. Among the insights fMRI has revealed are how trained tasters experience wine differently and how experience causes changes in the brain.

There is now hard science behind Goode's statement that "[k]nowledge of the molecular features that are being detected by the olfactory receptors is not sufficient to predict the nature of the smell that is perceived. How a receptor is 'read' by the brain depends on past experiences and current expectations." (p. 71)

Chapter 4 explains the chemistry that produces wine flavor. It is a handy reference for wine nerds interested in the names of polyphenols, wine faults, and other compounds. Here, sentences beginning like this: "Included in the contributory compounds are volatile phenols (guaiacol, eugenol, isoeugenol, 2,6-dimethoxyphenol, 4-allyl-2,6-dimenthoxyphenol); ethyl esters; fatty acids;…)" (p.88). While likely and admirably included for the sake of completeness by an author still enamored of his science, they can derail a nontechnical reader.

Chapter 5 examines "Individual Differences in Flavor Perception." The causes are discussed and illustrated in one of the many cartoon-like figures that break up, frequently repeat, and reinforce the text throughout the book. We learned that novices who lack depth of experience take a bottom-up approach to tasting, deriving their assessments from the sample itself. Professionals, who have accumulated experience and draw on considerable knowledge, use a top-down approach to form an opinion of a sample.

"Why We Like the Wines We Do" is the subject of Chapter 6. With both personal anecdotes and general observations, Goode describes how one begins to acquire new tastes. One study he cites emphasizes the importance of attentiveness to developing an affinity for an odor. "[A]ctively looking for [an aroma]… could reinforce the importance of having a wine vocabulary." (p. 140) This most significant observation might be enough for long-time tasters like me to return to taking notes more regularly, particularly when sampling the best bottles.

Goode, himself a professional wine judge, commendably maintains objectivity in considering whether wine expertise is real. While acknowledging the consistently poor performance of wine judges documented by Robert Hodgson (2008, 2009a, 2009b, 2017) and experiments that demonstrate how sommeliers can be fooled by the red dye in white wine, he argues that "[i]t only takes a few tasters—actually, just one—to taste blind accurately in any one situation to confirm that the skill exists…" (p. 142)

Chapter 7, "Constructing Reality," asserts that the brain needs a model. It develops an initial impression by predicting the responses of all the senses to stimuli. This impression is updated in reaction to deviations from expectations caused by the actual inputs. The updates continue until a reasonable model of reality is achieved that allows us to function.

Chapter 8, "The Language of Wine," discusses how we should communicate about wine. The experienced wine taster brings a lexicon of terms appropriate for each variety that serves as the basis for an initial model when he/she is about to taste. The model is updated to correct any discrepancies after the wine is tasted.

Goode shares his experience in building his own vocabulary and how he benefitted: "The words I had for wine were acting like pegs that I could hang my perception on, and because I was paying more attention…, I was seeing more." (p. 168) Yet Melanie McBride, a researcher at York University in Toronto, is given the last word in the chapter. She urges us to bypass words and concentrate on the actual sensory experience.

The penultimate chapter tackles whether or not wine tasting is subjective. The Goode doctor counters the pure subjectivists, who believe that all opinions are equal, by evoking the theoretical basis for wine assessment that people refine and absorb in order to attain recognized levels of expertise by passing tests.

"The assumption of those setting the exams is clearly that wine tasting, as carried out by professionals, is an objective practice." (p. 187) However, "the flavor of wine is not a property of the wine itself, but rather of the interaction between the wine and the taster" (p.188), Goode deduces from a statement by Gordon Shepherd. The degree to which wine tasting is subjective (though it appears that there is some evidence it isn't purely so) is still a matter of debate and research, with opinions coming in from both sides and in between.

What all this has been building to is the subject of Chapter 10. Goode exhorts us to "move away from thinking that we have five separate senses to seeing all sensation—and all perception—as just one sense." (p. 195).

In other words, take a "unite and conquer" rather than a reductionist approach of examining the pieces and ignoring the whole. There is no one right way to taste wine; how we taste depends on the objective. Wine professionals practice analytical tasting where the various attributes are identified and compared to prototypes previously learned. This requires a controlled, even sterile, environment.

On the other hand, "[i]f we are primarily seeking pleasure from our wine-drinking experience, we need to take an almost opposite approach to analytical tasting," (p. 199) Goode advises. In this case, the full context, environment, company, food, and music all contribute to the perception.

Goode classifies how a wine critic tastes as a third method though closely related to analytical tasting that must "extrapolate from the artificial setting…to the naturalistic setting in which [the] readers will be consuming wine" (p. 202). He leaves us with some good advice: "I am a firm believer in interrogating the wine…" (p. 203). He recommends training to improve our ability to do so. "If there is anything that can save objectivity in wine perception, it is that wine appreciation is largely learned, rather than innate," (p. 211) he concludes.

Goode has staked a position in a raging whirlpool of opinions that is being continuously fed by ongoing research. He supplemented his literature search with interviews of some of the major players. In casting a wide net, he uncovered interesting anecdotes and observations that provide amusing diversions.

A reference section at the end of the book lists some of the primary sources by chapter. A glossary gives definitions of many of the terms and a five-page, four-column index facilitates navigating the text. Goode's fluid writing augmented by cartoons elevated the book from what could otherwise, in places, be a pretty dry exposition.

Although the field of neuroenology is far from mature, enough has been learned over the past two decades that can be applied to heighten the tasting experience. Dr. Goode adopts a broad context for the state of knowledge and—unlike Dr. Shepherd whose recommendations are scant, less informed, and even naïve (Hulkower, 2017)—applies the current understanding to propose a sensible new model that should be tested by anyone aspiring to deeper appreciation of the most enthralling liquid. It is easy to understand how *I Taste Red* was named the Domaine Faiveley Wine Book of the Year at the 2017 Louis Roederer International Wine Writers' Awards.

References

Hodgson, R. T. (2008). An Examination of Judge Reliability at a major U.S. Wine Competition. *Journal of Wine Economics*, 3(2), 105-113. doi: 10.1017/S1931436100001152.

Hodgson, R. T. (2009a). An Analysis of the Concordance Among 13 U.S. Wine Competition. *Journal of Wine Economics*, 4(1), 1-9. doi: 10.1017/S1931436100000638.

Hodgson, R. T. (2009b). How Expert are "Expert" Wine Judges? *Journal of Wine Economics*, 4(2), 233-241. doi: 10.1017/S1931436100000821.

Hodgson, R. (2017). 2008 revisited: A review of judge performance at CSFCWC 2005-2012. Presentation at the 11th Annual Meeting of the American Association of Wine Economists, June 28-July 2, 2017. http://www.wine-economics.org/2017-padua/2017-padua-scientific-program/.

Hulkower, N. D. (2017). Review of Neuroenology: How the Brain Creates the Taste of Wine. *Journal of Wine Economics,* 12(3), 332-334. doi: 10.1017/jwe. 2017.34.

Shepherd, G. M. (2017). *Neuroenology: How the Brain Creates the Taste of Wine.* New York, NY: Columbia University Press.

Vineyards, Rocks, & Soils: The Wine Lover's Guide to Geology by Alex Maltman, New York, NY, Oxford University Press, 2018, 256 pp., ISBN: 978-0190863289 (hardback), $34.95.

E ver since I read Maltman's papers (Maltman 2013a, 2013b), I bristle when I hear the term "minerality" used in a description of a wine. Geologic minerals have no smell or taste, he insists. (We adopt Maltman's convention of using "geologic," what he calls the American convention, for tangible things and "geological" for mental constructs like time.)

How incredibly naïve it is, then, to think that a wine is a medium for transporting flavors from the land in which the vines that grew the grapes are planted. After all, when we smell flowers or taste berries in a wine, we know it isn't because it contains them. But, as others acknowledge: "Although many tasting terms are metaphorical…, there is a strong temptation to interpret 'mineral' rather more literally…" (Robinson 2015, p. 465).

Of course, not doing so could undermine a fundamental tenet of terroirists. Since the beginning of this century, "minerality" and "mineral" appear ubiquitously in wine writing. Maltman claims that "apparently it has now become the most widely used taste descriptor." (p. 176)

To me, it comes across as a vinous verbal tick that signals an indolent vagueness wrapped around a desire to flaunt a tuned-in palate. At times, I requested more specificity from visitors to a tasting room where I work when they claimed to have detected minerality, then sought validation from me, which of course, I never give.

So, I was amused and humbled as I was preparing this review when I read a tasting note that I had written while sampling a 1967 Chablis Grand Cru Vaudésir from Domaine Mary Drouhin in

1976: "Taste very flint and earth (sic)…Very earthy, minerally finish." (Hulkower 1976).

Oh, the irony! What was I thinking? Or more precisely, since I was still a novice, who or what was I channeling? The pervasive tasting note meme clearly has its roots going back many decades.

Since a long-practiced habit dies hard, a strong jolt is required to dislodge it. Maltman's excellent book is intended to be just that. The retired professor of earth sciences at Aberystwyth University in Wales and amateur vigneron deploys his formidable twin-pronged knowledge and pedagogical prowess in this volume aimed specifically at wine professionals, especially those who are perpetuating the myth of minerality in their writing. So ingrained is the idea that we can taste minerals in wine that numerous labels include the names of or references to rocks, minerals, and land features.

In the Preface, Maltman advises: "…these days it's almost obligatory in the wine world to know something about the geology of wine-producing areas and of particular vineyards." (p. xi) But he cautions: "…geology is a highly conceptual subject and not easy to pick up quickly." (p. xi) So, will those who are too lazy to be more precise in their descriptions be too lazy to read this challenging work?

One hopes that they are as receptive to Maltman's message as the celebrated British wine writer Andrew Jefford who, in the foreword, states: "…he is a scientist—and wine lover—with an open and enquiring mind who merely asks that we should understand what the technical terms mean before we use them and that we respect the journey toward understanding which science has so far permitted us." (p. x)

In the twelve chapters comprising the book, Maltman's approach is to teach us geology, literally, from below the ground up, starting at the atomic level with the elements (Chapter 1) that build the minerals (Chapters 2 and 3) that make the rocks (Chapter 4 to 8) that weather and erode and mix with biological material, called humus, to make the soil (Chapters 9 and 10).

Chapter 11, "Vineyards and the Mists of Geological Time," and probably the single most important section for the intended audience—

Chapter 12, "Epilogue: So is Vineyard Geology Important for Wine Tasting?"—complete the lessons.

Maltman reminds us: "…all rocks and soils are made from (geologic) minerals, not some more than others." (p. 173) Throughout the book, Maltman drives home the point that no geologic mineral can be sensed in a wine.

For example, regarding slate, which is prevalent in many vineyards, most famously in those on the Mosel and Rhine in Germany, he asserts: "…like most rocks, slate lacks any taste or odor. To have taste, a substance has to dissolve, and manifestly that is not the case with an inert material that makes practicable kitchen countertops and durable roofs." (p. 99)

There is another type of mineral, however—nutrient mineral—and therein lies some of the confusion. In addition to water, a vine only needs sunlight for photosynthesis and essential nutrients to thrive.

"Mycorrhizal fungi living in the soil can extract some [nutrients] directly from geologic minerals and transfer them into the vine's roots but otherwise complex weathering processes and ion exchange have to act to release the elements into the soil's pore water," (p. 167) Maltman explains. These nutrients are sometimes called mineral nutrients because they are extracted from the ground. But "most nutrition typically comes from the top few tens of centimeters or so of the soil," (p. 167) he notes. In particular, "the greater part of the nutrition comes from the organic matter in the soil." (p. 173) The critical process of cation (positive ion) exchange in soil water with the vine roots is masterfully explained in Chapter 2. Vine roots that grow deeper into bedrock are in search of water, not nutrients.

So, since vines absorb nutrient minerals but not geologic minerals, can we taste those? Well, for one thing, wines don't have much of them. "In normal wines, mineral nutrients typically comprise less than 0.2%, in total," Maltman informs us. Based on studies using water, a far less complex beverage than wine, "[i]t's possible that the tiny amounts can interact to produce some aggregate effect, but, tellingly, tasters report that as the presence of metal ions becomes increasingly detectable, the water becomes more and more disagreeable." (p. 177)

He concludes that "describing a wine as mineral or possessing minerality should not be referring to actual minerals—geologic or nutrient—but should be recalling some cue, some mental association." (p. 177)

For those willing to face the scientific facts but not all the details, the last chapter is a valuable summary and a firm persistent pushback on popular beliefs regarding the connection between the taste of wine and geology. The flavor of wine is largely created by our senses of smell and taste. "The taste components mainly involve ions and compounds in solution and geologic minerals are practically insoluble," (p. 217) Maltman reiterates.

Sodium chloride is an exception and gives a salty taste. But because "growers avoid salt in vineyard soils, and grapevines try to reject sodium…wine normally contains little salt, less than the minimum…most people require to be present *in water* in order to recognize it: a perception of saltiness in wine is usually metaphorical." (p. 217)

What is it then that is creating the impression that we are smelling and tasting rocks? Highly aromatic organic compounds like microorganisms are likely a source. A popular term in wine notes these days is "petrichor," the smell of rocks after a rain, which is caused by "the vaporization of certain organic oils (lipids, carotenoid, etc.)…" (p. 219).

Maltman addresses the iodine smell of the ocean in some chablis and makes the case that any iodine present would be in too small a concentration to be perceptible and "has to be a metaphor and unrelated in any direct way to the actual vineyard geology." (p. 219) Investigations are underway looking at bacteria lodged in the cleavages of minerals as a possible influence on wine, but nothing is clear yet.

Maltman has produced an important work that should give pause to those addicted to glibly tossing around "mineral" or "minerality" when referring to a wine's smell or taste. Though geology is a hard subject relying on its own sometimes confusing terminology and a bit of chemistry, his explanations—leavened with sly, wry, and even, once in a while, lame humor, as well as numerous charming digressions—are lucid. He draws from his deep and detailed knowledge of vineyards, wines, and wine-growing regions around the world to continually relate the geology to the interests of the intended readership.

Black-and-white illustrations mercifully break up the dense text but sometimes aren't sharp enough to highlight the features of interest. Fortunately, two dozen of them are also included as vivid color plates, albeit without the captions, so flipping back and forth is required. Most chapters conclude with an annotated list of suggested references. A six-page, two-column index assists the reader in finding a definition or first mention of a term and is essential in the absence of a glossary.

Despite all of the science refuting the notion of minerality in wine, I still perceive saltiness in a manzanilla or grower champagne, and chalk in a Pouilly-Fuissé. Is it real or is it a metaphor? Who am I going to believe—Maltman or my own palate? In the end, like Maltman, I accept that science must prevail and that, eventually, it will render these questions false dichotomies.

References

Hulkower, N. (1976) Tasting note for 1967 Chablis "Vaudésir," unpublished.

Maltman, A. (2013a). Mineral taste in wine, minerals in the vineyard…Are they connected? *Wines & Vines*, 94 (5), 63-70.

Maltman, A. (2013b). Minerality in wine: a geological perspective. *Journal of Wine Research*, 24 (3), 169-181.

Robinson, J. ed. (2015). *The Oxford Companion to Wine, Fourth Edition.* Oxford: Oxford University Press.

The use of the "m-word" in tasting notes is now so embedded in practice that it is as unlikely to dislodge as the word "the" in the wine critic's word palette.

Here is an amusing anecdote about an encounter with the m-word in the tasting room where I work. A customer asked me for a minerally wine. After somewhat less than diplomatically pointing out that this is a bullshit term with no universally accepted meaning, I asked what he meant by it. Untypically, he said, "Funky." Normally, I would expect the usual synonyms like "stony, earthy, salty, chalky, or even acidy." Without missing a beat, I poured for him a taste of a 2013 pinot noir from our older vineyard that, after almost ten years, was just harmonizing an extreme funkiness that could have been caused by Brettanomyces or Brett, a yeast found in many wineries. He was thrilled and purchased a bottle of it as well as a 2011 from the same vineyard and another 2013 from a neighboring vineyard. Hey, whatever works and makes a customer happy!

Wine Girl: The Obstacles, Humiliations, and Triumphs of America's Youngest Sommelier by Victoria James, New York, NY, Ecco, 2020, 336 pp., ISBN: 978-0-06-296167-9 (hardback), $26.99.

A memoir by someone under thirty? Outrageous! What can we learn from someone so young, especially if thirty is the new eighteen? Quite a bit, actually.

In less than a decade, Victoria James has gained important recognition. She became a Certified Sommelier in the Court of Master Sommeliers at age twenty-one. In 2013, she won the Chilean Wine Challenge in New York and was named Best Sommelier of the Sud de France. Two years later, she came in first in the Ruinart Champagne Sommelier Challenge in New York, was named by *Forbes* as one of the "Top 10 Innovators under 30 in New York City," and earned a place on Zagat's "30 under 30" list. She made *Wine Enthusiast*'s 2016 "Top 40 under 40 Tastemakers" list and secured the title of *Wine & Spirits* magazine's "2016 Best Sommelier."

In 2017, she published her first book, *Drink Pink: A Celebration of Rosé*, which was illustrated by her future husband, Lyle Railsback. *Forbes* included her on its 2018 "30 Under 30" list. That same year, *Food & Wine* called James the "Best Sommelier in New York City."

She is now a partner and beverage director at Cote in New York City and cofounder of Wine Empowered, a nonprofit helping women and minorities enter the hospitality industry. But what makes her story so special is what she had to overcome to finally succeed and how she did so.

The prologue describes a disturbing encounter with some obnoxious customers, involving a bottle of 2009 Domaine Ramonet Chevalier-Montrachet, during James's first year as the youngest sommelier in the US, at which time she was first called "wine girl." She then presents her story in seven parts, each covering a span of ages: 7–14, 14–19, 19–20, 21–23, 23–24, 24–26, and 26–28. Each part contains three or four chapters with titles including "The Poor Kids," "The Wine School," "Wine is a Blood Sport," "Watch Out for the Wives," and "The Real World."

In Part I, James, the second of four children, recounts her childhood from hell. "I grew up in a household of manipulation and neglect, left to fend for myself," (p. 242) she stresses. Her mother was an Italian countess but lacked the basic skills needed to be an effective spouse and parent. Her father came from a humble background, had strong analytic skills that he never seemed to apply, and was frequently absent and broke. He eventually fell into alcoholism and developed a gambling addiction.

"This juxtaposition of blue blood and blue collar is what I believe groomed me for the eventual role of a sommelier, essentially a high-brow servant," (p. 11) she explains. Her mother homeschooled her brood until succumbing to frailty, an overbearing husband, and the demands of raising her children. She became reclusive.

With their father off on extended trips, the children were left to fend for themselves and had to cobble together meals from what few edibles were around. "We had to make the sleeve of crackers last for days," (p. 16) she recalls.

James's first foray into the beverage business was a lemonade stand. Her father collected the cost of materials and a rental fee for the table. Becoming a better shopper and more attentive to customer preferences led her to succeed in her first beverage director experience.

A particularly traumatic episode involved eight-year-old Victoria slamming the car door on her mother, as James's father, fed up with his wife's neglect, took the children to live with his mother in their aunt's house. "The Poor Kids" covers the time of the divorce proceedings, tutoring by an unemployed, physically and mentally abusive father, and little money. It also describes the transition to public school where the children skipped a grade after testing high. "I found that because my father had drilled into me with relentless focus, I always forced myself to learn at an unusually accelerated pace," (p. 34), James acknowledges.

At age thirteen, James was hired as a waitress at a diner in New Jersey where her father landed them. Part I concludes with the story of her friendship with the cook and dishwasher. Despite the grossness of many of the tasks, the two never grumbled. "Sometimes the more honest the work, the more honest the people," (p. 51) observes James.

During her teens, the era comprising Part II, new challenges and traumas arose. James's father returned from triple-bypass surgery with a new partner who gave him a fourth daughter. Later, he succumbed to alcoholism and gambling. During the frequent trips to Atlantic City so her father could indulge his vices, James kept busy by becoming an "unofficial drink runner" (p. 64), collecting chips as tips and learning about classic cocktails along the way.

James took a job at the Plaza Diner when she was fifteen and credits Franky, the most senior waiter, with teaching her how to be a "real hospitality professional." (pp. 68–69) "The advice I most appreciated from Franky was about customers. 'You gotta love 'em, I mean it…Tell 'em you love 'em, too,'" she remembers. (p. 69) While this advice generally served her well, it led to her rape by a regular during the graveyard shift. Because he seemed trustworthy, she accepted a ride home. He attacked her in his car not far from her house. She did not discuss the trauma with anyone and suffered nightmares. She also began to abuse alcohol and drugs.

Though Part III is titled "Age 19–20," it really starts earlier. At seventeen, James was in New York in college on a scholarship. Her academic career ended in less than a year as she continued to indulge in booze and drugs. An aunt took her to California, but this did not turn out well, and she ended up back in New Jersey living with her father. A few months later, having kicked drugs with the help of therapy, she returned to New York City where she has been since.

In the Big Apple, James took a series of positions in higher-end restaurants in which she expanded her knowledge of cocktails and developed a love of wine. At Lattanzi, she had her first wine-and-food-pairing epiphanies. She enrolled in a wine school where she met her first sommelier. At Harry's, she acquired a mentor who shared some of the treasures of its legendary cellar, though not with permission of the owner. There, she assisted in reorganizing the collection when not tending bar. But after teaching her about wine and introducing her to haute cuisine, the mentor began to pressure her. "I so desperately wanted to be in that wine cellar, to learn and touch all those rare bottlings. I needed to make a living, too. So eventually, I gave in," (p. 134) she admits.

At the recommendation of her mentor, James worked a harvest in Sonoma. When she returned to New York City, she began to see an acupuncturist/therapist to deal with some of the physical effects of the harvest. James also credits her with helping her cope with her emotional issues.

Things started to look up when James was twenty-one to twenty-three, the period covered in Part IV. With the help of the instructor at the wine school, James became a sommelier at the Michelin-starred Aureole. She continued at the wine school, entered wine competitions, and became a Certified Sommelier. She later migrated to Marea, a Michelin two-star restaurant.

But as James discovered, Marea was but a way station to Ristorante Morini, a new establishment she would help open. Part V, covering "Age 23–24" details the rise and demise of that venture following a bad review by *New York Times* critic, Pete Wells. She returned to Marea and became a full-time sommelier after surviving a grueling trial period.

Part VI, "Age 24–26," describes James's transition from a hectic and toxic environment to an island of tranquility. During this time, James took trips to wine regions out of the country that she had won in competitions. There was also recognition in several publications as noted above.

Despite her growing reputation, James lost her job at Medea the day after she learned she was on Zagat's "30 under 30" list. But there was also another rape, this time by an unnamed boss in a wine cellar. The abuse to which she acquiesced for fear of losing her job lasted months.

Now shuttered, Piora became James's next opportunity. It brought serenity and respect into her working life, and when it closed, an even greater chance to grow and excel. It too had been Michelin-starred under the proprietorship of Simon Kim, who not only took care of patrons but also the small staff. There James thrived.

As wine director, she "instituted a policy where a guest could order *any* bottle on the list and only commit to drinking half of it, hence paying half price." (p. 254) She also was responsible for purchasing wine. Unlike many restaurants in New York City that mark up wine three to four times retail, James used a factor of 2.9 and sold twice as many bottles.

At the end of Part VI, James shares the story of how she met her husband, Lyle, a wine salesman for Kermit Lynch. The final part, "Age 26–28," continues with reminiscences of the courtship and ends with the wedding at the family castle in Piemonte. It also describes the closing of Piora and the opening of Cote, the Korean steakhouse awarded a Michelin star a few months after. There James insists that all wines by the glass are poured from magnums bottled especially at the sources in France because "the wine…stays fresher for longer and tastes better." (p. 289)

While at times utterly depressing and infuriating, *Wine Girl* is an engaging read. James's style is forthright and unadorned. It must have been painful for her to recount the abuse she experienced or witnessed, which she does in explicit language. Admittedly, I was hoping for some occasional respite from the unalloyed descriptions of the mistreatment she endured. A little more on notable wines she had tasted, food pairings she recommends, or positive encounters with patrons would have helped leaven her heavy narrative. Nevertheless, I came away in awe of this young star whose intelligence, persistence, and resilience are bringing her recognition and peace. "I had grit," (p. 220) she proudly declares.

Wine Girl is the "wine cellar confidential" in the age of the Me Too movement worthy of our regard. James's cathartic account of her physical and psychological abuses and indignities over the recent past should serve as a cautionary tale, reminding us that despite increased opportunities for women in male-dominated professions, attitudes and behaviors remain largely unchanged. A female coworker's response to a customer who was "calling her the most horrid names" (p. 195) bears repeating: "'Sir, with all due respect, I am a human being. I would demand, at the very least, recognition as such.'" (p. 195) James's poignant story needs to be told to everyone in the hospitality industry and to those who come in contact with it. In other words, everyone.

Pinot Girl: A Family. A Region. An Industry by Anna Maria
Ponzi, Sherwood, Oregon, Bristol Press, 2020, 372 pp., ISBN:
978-1-7345788-0-5 (paperback), $17.95.

Unlike previous accounts of what it was like to be there at
the beginning of the wine industry in the Willamette Valley,
Maria Ponzi's memoir is from the point of view of a child.
Works such as those by Susan Sokol Blosser (*At Home in the Vineyard:
Cultivating a Winery, An Industry, and a Life; The Vineyard Years: A
Memoir with Recipes*) tell the story from the perspective of the adults
who made the decision to relocate to Oregon to grow grapes and
make wine in a region deemed inhospitable for this crop as opposed
to someone who was swept along.

When the success of the region began to be recognized, stories of
the pioneers became the subject of several books from an outsider's
perspective written by industry observers including Vivian Perry
and John Vincent (*Winemakers of the Willamette Valley: Pioneering
Vintners from Oregon's Wine Country*), and Cila Warncke (*Oregon
Wine Pioneers*).

As the first generation of the industry founders retire, the issue
of succession is being resolved in different ways, with a few of the
original wineries passing to the second generation. Ponzi's well-told
tale recounts her initial resentment toward being expected to work
from an early age, her brief foray into a different industry, and finally
her return to embrace the family business wholeheartedly and assume
its leadership.

Ponzi acknowledges in the introduction: "At first glance, a young
girl born into the wine business may seem like she has a charmed
life. While the romance has slowly evolved, it was far from that in
my early years." (p. xi) Her story is told in two parts, each comprised
of short chapters.

Part One, by far the longer, covers the period from her birth in California through her marriage in 1995, about four years after her return to Oregon. Events of the past couple of decades are very briefly summarized in Part Two. The one-page epilogue is a snapshot of the current state of Ponzi Vineyards and the Oregon wine industry. An album of black-and-white photos of the family, from 1968 to 2019, adds a valuable visual dimension to the history.

Born in 1965 in Los Gatos, California—the same year David Lett, known as Papa Pinot, planted the first *Vitis vinifera* vines in the Willamette Valley—Ponzi had a front-row seat during the entire birth and growth of the wine industry. Her parents, Dick and Nancy Ponzi, are among the venerated dozen or so first families who took the leap to make wine where few thought it possible and succeeded beyond their expectations.

During a trip to visit his older brother in Iceland in 1967, Dick tasted a homemade wine fermented from a vegetable. Dick and his brother reminisced about their father's winemaking, which was done without protecting against oxidation. "This conversation and the delicious celery wine sparked something in my father that night. He left the island inspired," (p. 13) reports Ponzi.

Back home in California, Dick began making wine from grapes he picked at the Novitiate Winery. Two-year-old Maria and her four-year-old brother, Michel, participated in the harvest and crush along with their pregnant mother. Luisa Ponzi, destined to become the winemaker of the family business, was born two months later.

The family moved to Oregon the following year after a visit to Nancy's parents who had retired to the Willamette Valley. During that trip, Dick met Charles Coury who, along with David Lett, is credited with bringing pinot noir to the valley in the late 1960s with plantings that produce to this day. "It was after that meeting in Forest Grove when everything seemed to fall into place for my parents," (p. 23) states Ponzi.

Moving his family into a 700-square-foot shack on a strawberry farm in Scholls, Dick traded his engineering career in California for the life of a winegrower and farmer. To fund his enterprise, Dick taught mechanical engineering at Portland Community College while Nancy ran the household. Taking cues from the neighboring farmers, the Ponzis figured out when to plant.

"Instead of taking classes to learn this stuff, they'd just pick up another book and read," (p. 38) reveals Ponzi. Though their dreams were met with derision and disapproval, the entire family proceeded undeterred to plant their first vineyard in the spring of 1970. Soon after, crops were planted and animals were brought in to provide food for the family. And as if there wasn't enough on his plate, Dick built a house while Nancy tended the vineyard.

"Each day they'd put in long hours, pushing themselves until… the darkness would…force them to put down their tools…We rarely had dinner before nine-thirty…" (p. 95) remembers Ponzi.

During the 1970s, the vineyard and farm matured, the first grapes were harvested and made into wine, the house and winery were completed, and the children began school. Maria coped with the difficulty of fitting in with children who echoed their parents' opinion of the hippies from California trying to grow grapes.

It was also the time the "Pinot Obsessed," as Ponzi calls the first winemakers to plant the finicky grape in the Willamette Valley, came together. Myron Redford, Dick Erath, Coury, and David Lett were part of "the group of young winemakers [who shared] a mutual love for the Pinot Noir grape and [dreamt] of what could be" (p. 111). In addition to sharing insights, tools, labor, and resources, some members lobbied the Oregon legislature to pass laws to protect the budding wine industry.

Since money was short in the early days, trading wine for goods and services provided relief. "The five of us were well cared for by a local dentist for years through several cases of Riesling," (p. 222) boasts Ponzi.

Recognition of the quality of Oregon pinot noir, in general, and Ponzi's, in particular, came quickly. The Eyrie Vineyards South Block Reserve 1975 made by David Lett placed second among pinot noirs and above all but one burgundy in the 1979 Gault-Millau Wine Olympics. The Ponzi Vineyards Pinot Noir 1977 impressed Frank Prial of the *New York Times*. "Not only did he write about it, but he *loved* it!" (p. 230) effuses Ponzi.

Tensions with her mother increased when Ponzi entered high school. Nancy disapproved of any activities including cheerleading and typing that ran counter to her feminist perspective. "Instead, she encouraged my interests in writing and politics," (p. 231) she explains. At the same time, she gained acceptance by her classmates. Still, she was expected to help around the vineyard and "did [her] best to sneak out of work." (p. 234) "When I did complain, reminding Mom that 'I didn't ask for this life,' her standard response was, 'Well, this is what we're doing, and that's the way it is,'" (p. 235) Ponzi laments.

After college, Ponzi took a job in advertising in Boston. From the East Coast, she learned of the latest praise for her father's winemaking. After three years, she was offered a big promotion at the same time as the news from home intrigued her. "I was curious about the new-comers to the valley and suddenly felt territorial. I realized I missed my parents and the bustling family activities," (p. 280) she admits. "I yearned for something more. I felt perhaps the 'grown-up' winery could provide that for me." (p. 281) She returned to Oregon in 1991 and dove into marketing.

Twenty years are covered in the nineteen-page Part Two. Significant events include the birth of her children, the continued recognition of Dick Ponzi in the wine press and by his peers, the ascendance of pinot noir in the wake of the movie *Sideways*, and the construction of a new winery.

I found just a couple of minor discrepancies. On p. 35, we read: "In 1965, Lett planted the valley's first nursery in McMinnville…," yet according to the winery's website (https://eyrievineyards.com/envisioning.shtml): "In February 1965, David [Lett] rented a temporary nursery plot near Corvallis, and planted the 3000 vinifera grape cuttings he gathered from UC Davis and selected growers and brought with him to Oregon. This was the first planting of Pinot noir and Chardonnay in the Willamette Valley, and the first 'New World' Pinot gris."

Also, there may be conflicting information. On p. 149, we are told: "By 1975, there were fourteen wineries in Oregon," but on p. 200: "By now there were nearly ten wineries in the state." Since the story unfolds chronologically, this gives the impression that the industry was contracting. Was it?

Ponzi's narrative is interwoven with many quotations either attributed to herself or recalled by her. This is remarkable given that she was as young as two years old when the events she describes were taking place. As in the last movement of Mahler's Symphony No. 4 when the soprano sings *"Das himmlische Leben,"* it seems like we are hearing the voice of a child at times, though it certainly was not recounting a heavenly life. "[This intimate tale] draws almost entirely from childhood memories, keen eavesdropping, and lengthy tableside conversations" (p. xi) asserts Ponzi. She also interviewed some of the pioneers and did additional research, but "most of this story draws from my recollections…Like a child's view, it is intimate, honest, and pure," (p. xii) she says. In later chapters and in Part Two, it also seems that the tone of the writing matures along with Ponzi.

With the release timed to celebrate the fiftieth anniversary of Ponzi Vineyards, *Pinot Girl* is a valuable addition to the literature recounting the birth of one of the most exciting wine regions in the world less than sixty years ago. Ponzi's deeply personal and detailed account offers insights into the evolution of her attitude toward the family business, at each stage of its development, which is distinctive from those of the first generation and of wine writers and journalists. It is also a window into how one person came to embrace her destiny to become part of the second-generation leadership in one of the first wineries in the Willamette Valley. Read it.

The second-generation takeover of the winery lasted thirty years. In 2021, Groupe Bollinger purchased Ponzi Vineyards, their first winery purchase outside of France. Included in the sale was the winery, tasting room, and some of the vineyards, the majority of which remain with the Ponzi family. Luisa Ponzi is currently listed as Legacy Winemaker on the website. Maria serves as the director of Linfield University's Wine Studies Program.

The Story of Wine: From Noah to Now, New Edition by Hugh Johnson, Académie du Vin Library Ltd., 2020, 496 pp., ISBN: 978-1-913141-06-6 (flexibound), $45.

R ule Britannia! Britannia ruled the wine world in Europe, at least between the eighteenth century and well into the twentieth. This is all the more remarkable since only until recently has the UK produced anything vinous worth mentioning, and not all that much at that.

Even more ironically, as an example of always hurting the one you love, a vine disease, *oidium tuckeri*, passed from England to France via Belgium in 1851 and wreaked havoc for a decade until it was found that dusting with sulfur would stop its spread.

England's more enduring positive impact was as a thirsty market for the wines of France, Spain, Germany, and Portugal. Hence, it was the most important stimulus for elevating their quality. As Hugh Johnson affirms: "Great wines are made by their markets." (p. 245)

Alongside the growing wine trade, wine writing began to flourish in Britain starting in 1775 with the publication of *Observations, Historical, Critical and Medieval, on the Wines of the Ancients and the Analogy between Them and Modern Wines* by Sir Edward Barry.

"Writing about wine from the consumer's point of view had in the past been almost a branch of medicine…" (p. 317) explains Johnson. "Such writing was to become the specialty of the English, for the simple reason that English wealth, at the top of the social ladder, had accumulated the most varied cellars of top-quality wines on earth." (p. 317) Building on this tradition, Johnson presents his masterful overview of the evolution of the world's most cherished beverage, not surprisingly in an unabashedly Anglocentric way.

Following the foreword by Andrew Roberts (more on this below), Johnson's preface addresses his wariness of the word "history" and the limited degree to which he updated the story after thirty years. On the former, he recognizes that "scholars have made [the word history] their own and will challenge any unqualified pretender." (p. 11) His reluctance to go on with the story stems from his belief that it would be "as much about money as it is about wine, or taste or pleasure." (p. 14) Moreover, he asks: "It would further our knowledge—but does it further our understanding? In any case…I have limited myself to the story of wine at the time when it took over a large part of my life, and became my enduring pleasure." (p. 15)

Gustave Flaubert declared that "Writing history is like drinking an ocean and pissing a cupful." Having eschewed the "h" word, Johnson instead—after consuming thousands of wines, including one that was 421 years old and digesting dozens of references—penned his tale that unfolds across forty-three chapters organized into five parts. The nine chapters of Part One cover the period from "Man's First Experience of Alcohol" to when "Mohammed Condemns Wine." (I'm using chapter subtitles which are more descriptive of the content than the titles.)

Part Two, with ten chapters, begins with "Charlemagne and the Rebirth of Europe's Vineyards" and ends with "Great Steps in the Technology of Glassmaking."

The eleven chapters of Part Three cover the period from the end of the seventeenth century into the nineteenth. Part Four comprises ten chapters that dwell on the turbulence caused by revolution and wars. They address their impact on the maturation of the wine industry, primarily in Europe but with excursions to the New World, and take us to the dawn of the twentieth century.

The first of three chapters of Part Five is an overview of the first half of the twentieth century, featuring "War, Slump, Poor Weather and Prohibition" and followed by two containing updates since the first release of the book. Up-to-date information is also spliced into the text throughout, for example: "…the oldest pips of cultivated vines…were found in Soviet [as it was then] Georgia…" (p. 24)

Mercifully, an index of nineteen pages, each containing three columns of smaller text, is included to facilitate refreshing the reader's memory. A bibliography lists references by chapter. Unfortunately, there are no maps or illustrations.

Freed from the strictures of historians, Johnson takes a totally relaxed and occasionally personal approach to his subject, covering many topics superficially while going down the rabbit hole on others. He infuses the text with opinions, wit, and amusing digressions but not at the expense of a serious message. This story shows how much wine permeates and influences various cultures, and so must include some of its history. Wine is overlaid on this chronicle with names and places and empires and raiders poking through the tales, requiring the most curious to look elsewhere for details.

Although I am well-read on the subject of wine, I learned quite a lot. In particular, insets were especially helpful to break up the formidable blocks of text and further illuminate a topic. Some offer quirky insights. For example, one named "Tent" introduces a term no longer used for a dark red wine and "is the directly comparable with, and complementary to, 'claret.'" (p. 167) Johnson muses: "It would be pleasant to see it introduced for that general class of wine such as Australian Shiraz, California Zinfandel, and indeed such dark Spanish reds as Duero (as opposed to paler Rioja)." (p. 167)

As another example of the myriad topics touched on Johnson tells us that "Plato's views on the minimum drinking age are remarkably severe. 'Boys under 18 shall not taste wine at all for one should not conduct fire to fire. Wine in moderation may be tasted until one is 30…But when a man is entering his fortieth year…he may summon the other gods and particularly call upon Dionysius to join the old men's holy rite…wine…is the cure of crabbiness of old age…' It is a sobering thought that to Plato old age began at 40." (p. 50)

Johnson's remembrance of the oldest wine he tasted opens Chapter 29, "Cabinet Wine." In 1961, he sampled an 1857 Rüdesheimer and an 1820 Scholl Johannisberger: "Both had completely perished…But the Steinwein of 1540 was still alive. Nothing has ever demonstrated to me…that wine is indeed a living organism…It even hinted…of its German origins." (p. 288) The exposure to air quickly turned the ancient liquid into vinegar. "It was a moving event in any case to drink history like this," (p. 289) Johnson concludes.

Johnson isn't only one of the most prolific wine writers, but he is also one of the most literary. Chapter 33, "Methode Champenoise," begins with this description of a scene shortly after the abdication of Napoleon:

"The sun rising over Champagne on the September 10th 1815 found something more stirring to illuminate than the usual placid dewy vines, their leaves yellowing and their grapes turning old for the approaching harvest…where the first light had touched the little hill…, a seemingly endless army was assembling…The light of dawn flashed on the cuirasses of hussars and glowed on the bearskins of great-coated grenadiers." (p. 332)

Passages like this are not only a pleasure to read but help the reader conjure up mental images that might partially offset the lack of illustrations.

At the same time, while impressed at the sheer scope and depth of the coverage, I feel that more could have been added to the update despite Johnson's reasoning. There are only three references to China listed in the index. The first reduces the early events in that country to a single page inset entitled "Far Cathay" (p. 28). The second is a reference to tea, not wine, and the third refers to the number of bottles packed in a hamper sent to China. Surely, room could have been made in Chapter 42, "New World Challenges," for an overview of the burgeoning wine industry in one of the world's largest wine markets.

In the foreword, Roberts acknowledges that "No one could be better qualified to write the story of wine than Hugh Johnson whose name is synonymous with wine writing." (p. 7) True enough. "There is tremendous scholarship to be found in these pages, but the immense learning is never ponderous. It is erudite, but never pompous," (p. 7) he maintains.

Also true, but it is not light reading. Readers should have at least an intermediate level of wine knowledge as well as more than a passing understanding of world, especially European, history and geography.

Without the slightest hesitation, Johnson casually references wines and wineries as well as major events throughout history, assuming that they are common knowledge. Admittedly, it is on the reader to fill in the blanks, but doing so while engaging with the dense text can be distracting and break the flow of the story.

Nevertheless, armed with a rare depth of knowledge and understanding, Johnson admirably, skillfully, and literarily undertook the formidable task of summarizing the development and expansion of viticulture and enology since biblical times. Particularly in the insets, he larded and leavened his saga with enough trivia, facts, and factoids to make the reader, depending on the guests, either a genius or a bore at the next dinner party.

But despite its title, this is only one story of wine, obviously limited to the interests and perspectives of the author, as he admits, and one that is distinctly Eurocentric. Johnson's rationale: "I was drawing a line between regions where wine had evolved through history and regions where it was an import based on what was being done elsewhere—almost all in Europe." (p. 14)

While that continent is where a lot of the action has been happening, it is certainly not the only one, especially since the original publication date. Perhaps there is a nonEuropean counterpart to Johnson who will pick up the story someday and celebrate the contributions of those in less obvious but equally important places.

Hugh Johnson sent me the following email: "I must sincerely thank you for your careful and evidently caring review. I have only just met it. In future, shall add this excellent journal [Journal of Wine Economics] to my pile.

"I do take quite a lot of stick for writing as an Englishman, but trying to adopt any other persona would convince few, and besides would slow me down. I do hope one of your countrymen takes up your final challenge. You, perhaps?"

While I appreciate the sentiment, I don't plan on embarking on such a project. I'm perfectly content just to incorporate bits of

*wine history in my pieces and not attempt to do an entire survey.
Johnson's extraordinary effort is an example of how fraught such
an undertaking would be. And then there are critics like me
waiting to nitpick. No thanks.*

In Vino Veritas: A Collection of Fine Wine Writing Past and Present edited by Susan Keevil, Académie du Vin Library Ltd., 2019, 224 pp., ISBN: 978-1-913141-03-5 (hardback), £30.

Okay, I'll warn you upfront: I'm enamored with British wine writing. The dry wit, the masterful yet effortless use of language, and the confident command of the subject remain inspirational models for this wine writer.

In Vino Veritas assembles thirty-six pieces, dating from 1833 through 2019, by thirty-four writers, many of whom are English, including Michael Broadbent, Hugh Johnson, and Steven Spurrier. The latter two, along with Simon McMurtrie, founded the Académie du Vin Library, the publisher of this volume. A brief introduction by Hugh Johnson, highlighting the origins of wine writing, is followed by ten chapters, each covering a single theme and each containing three-to-five stories, each of which is two-to-eleven pages long.

Charles Walter Berry's "In Search of Bordeaux," is a highlight of Chapter 1, "Good Vintage, Bad Vintage." This excerpt, from *In Search of Wine, A Tour of the Vineyards of France,* published in 1935 by "one of the first British wine merchants to venture abroad and taste wines on their own terroir" (p. 18), chronicles his visits to the chateaux the year before.

It contains descriptions of wines, both good and not-so-good, and accompanying dishes. Included as an insert, Michael Broadbent's tasting notes of several of the wines supplement Berry's pithier ones. In addition, Fiona Morrison MW contributes "Le Pin: the First Day of the Harvest," written in 2019, and H. Warner Allen describes "My Best Claret" (1951).

A 1981 extract from *Christie's Wine Companion* by Michael Broadbent, "My Wife and Hard Wines," concludes the chapter. It is a charming recollection of visiting old wine cellars whose bottles ended up on the block at the famous auction house. Contrasting "map-bedecked modern American air-conditioned cellars" with "the 'feel', smell, chill and content of an old cellar," he wonders: "How can a room comfortable enough to sit in for several hours…possibly be the right temperature for storing fine vintage wines?" (p. 25)

"Bordeaux, Burgundy…or Napa Cabernet?" is the focus of Chapter 2. It starts with a debate, of sorts: "Burgundy is Better" (1940) by Maurice Healy versus Ian Maxwell Campbell's "Burgundy, The Cannibal Wine" (1945).

Spurrier's "The 'Judgment of Paris' Revisited," written in 2018, details the results of subsequent rematches of the 1976 tasting he organized, as well as the original event. He quotes Ashenfelter and Quandt (1999) who "concluded that: 'It was no mistake for Steven Spurrier to declare the California Cabernet the winner.'" (p. 41) This is true based on the inclusion of the rankings of Spurrier and Patricia Gallagher in their analysis. However, there is strong evidence that their ratings were not included (Taber 2006, Hulkower 2009).

As I demonstrated, without Gallagher's and Spurrier's points, top honors went to the 1970 Château Haut-Brion. Nevertheless, this article is a valuable record of a tasting that "gave to the world of wine…a template whereby little-known wines of quality could be tasted blind against known wines of quality…" (p. 45). I can drink to that.

"Power to the Underdogs," Chapter 3, includes "Notes on a Barbaric Auslese" (1920) by George Saintsbury, "The Debut of Dom Pérignon" by Henry Vizetelly (1879), and a 2019 philosophical musing on an obscure variant of Syrah called Sérine, "Ah, the Sérinity…," by one of the original American Rhône Rangers, Randall Grahm. Also from 2019 is a credible analysis of the future of British sparkling wine, "The English Wine Bubble" by Justin Howard-Sneyd MW. He cautions:

> "When there is not enough wine to go round, no producer ever needs to price-promote, and no retailer wants to create a price war…This rather artificial environment trains the customer

to pay the full price and buy the wine…immediately…But this state of the market can quickly unravel as soon as supply exceeds demand, even by a small amount…it looks as if this is where English Sparkling wine may be headed next." (p. 73)

Kathleen Burk's 2013 article, "Cyril Ray and The Rise of The 'Compleat Imbiber,'" is a delightful short history of the publication that inspired this volume.

Chapter 4, "Wine Travels," comprises four accounts, either first- or second-hand, of visits to regions around the world. Hugh Johnson goes to "The Wilder Shores of Wine" (2019), Peter Vinding-Diers is "A Viking in the Vineyard" (2019); Simon Loftus spends time with "Guiseppe Poggio: Home Winemaking in Piedmont" (1986); and Jason Tesauro extols the progress made in winemaking in Virginia in "Out of California's Shadow" (2019).

Three-piece Chapter 5, "The Mischief of Tea," follows, offering quirky views of the English staple, especially vis-à-vis alcohol from George Orwell (1946), Cecil Torr (1918), and PG Wodehouse (1964).

Chapter 6 is a four-way discussion across 186 years over the question: "Should Port be Fortified?" The title, "A Call to Ban Port's Fortification," (1833) by Cyrus Redding, unambiguously stakes out one position. Dirk Niepoort's "The Best of Both Worlds?" (2019) defends the middle ground, which in fact, is put into practice at the company bearing the family name. An excerpt from *A Contemplation of Wine* by H. Warner Allen (1951) examines "The Scandal of Elderberries," involving adding the juice of this fruit to darken port. The final word, which I leave to the reader to discover, is given to Ben Howkins in "The Port Trials" (2019).

"To the Table at Last," Chapter 7, is a quartet of essays by three Brits and one of the most distinguished American wine writers of the last century. Jane MacQuitty ponders, "To Decant or Not to Decant?" (2019). Hugh Johnson's "Beyond the Banyan Tree" (1980) is a remembrance of a dinner organized by the Zinfandel Club, during which a selection of notable California vintages was served.

Californian Gerald Asher recounts the challenges of serving the best from our collections at a multicourse dinner in "Wine on Wine" (1996). As in Berry's story, Broadbent's tasting notes of some of the clarets and California cabernets are included. Spurrier's "Memorable Menus" (2019) will leave the reader both envious and incredulous as to how anyone could consume that much and still live to write about it.

Chapter 8, "Something a Little Different," is the shortest, with just two very brief essays. "Sting Like a Bee" (2019) by Dan Keeling considers high-alcohol wines. Jonathan Miles offers "Mint Julep, A Cocktail to Crave" (2008) as his complete departure from the subject of the book.

Chapter 9 looks at "Wine and Art" from four perspectives. Australian Andrew Caillard MW, who is also a painter, explores "Art, Wine and Me" (2019). The editor's introduction, of course, includes the inevitable wordplay: "The palate and the palette have been tools of his trade for over 40 years." (p. 166)

American wine writer Elin McCoy contemplates "Is Wine Art" (2018) and reveals a third side to this coin. The backstory of the decision to put art on the labels of Château Mouton Rothschild is disclosed in "Best Dressed and Bottled at Home" (1984), extracted from *Milady Vine, the Autobiography of Philippe de Rothschild* by Joan Littlewood. Canadian Tony Aspler writes about one Mouton label in "For a Piece of the Glamour" (1997) with an ending that could come straight from O. Henry.

A trio of essays comprises the final chapter, "Wine and the Poets." Baudelaire is the focus of "Wine and the Outcast Poet" (2009) by Giles MacDonogh. "Colette and Wine" (1983) by Alice Wooledge Salmon offers vinously inflected highlights of the life of a talented but controversial character. We are cautioned that "just as one's pleasure in rare wine can be blunted by undue dissection, so various critics have taken Colette to absurdities in their haste after 'psychoanalysis' of both woman and achievement." (p. 211) In any case, one can certainly lust after some of the bottles she encountered. Harry Eyres discusses Roman poet Horace in "In Vino Veritas" (2014).

In Vino Veritas is a gorgeous package, stunningly illustrated with exquisite color and historic black-and-white photographs and handsomely bound with a blue ribbon marker. I did find a few minor editing issues, however. On p. 46, "vineyard" has a typo; on p. 66, "to this" is repeated; and a caption refers to its illustration in the wrong direction (p. 85). Nits to be sure, but unexpected, given how carefully the book was otherwise compiled.

This assemblage of small sips, tantalizing tastes, and gratifying gulps of some of the best wine writing of the last two centuries is a joy to read. It is as much a verbal crazy quilt as an anthology that is clearly self-aware, with footnotes referencing other essays and an occasional piece responding to another in the collection.

I reveled in the precision of the jewel box of $50 words (€41.29 or £35.44 on June 11, 2021) like "cacographists" and "omnibibulosity" (p. 17), "etiolated" (p. 67), "adventitious" (p. 124), "flagitious" (p. 127), and "topos" (p. 221), as well as the French "maquillage" (p. 62), all of which I had to look up.

I really didn't want the book to end. Good news, signaled by the date 2020 on the spine and front cover, was found on p. 224 buried in the Acknowledgments: "With luck and a following readership, perhaps our book [will] see a run of annual editions…" Sign me up.

References

Ashenfelter, O. and Quandt, R. (1999). Analyzing a Wine Tasting Statistically. *Chance*, 12, 16-20.

Hulkower, N. (2009). The Judgment of Paris According to Borda. *Journal of Wine Research*, 20 (3), 171-182.

Taber, G. (2006). *The Judgment of Paris: California vs. France and the Historic 1976 Paris Tasting That Revolutionized Wine*. New York: Scribner.

But alas there has yet to be another collection.

The Science of Wine from Vine to Glass, 3rd Edition by
Jamie Goode, Oakland, University of California Press, 2021,
224 pp., ISBN: 978-0-520-37950-3 (hardback), $39.95.

A s evidenced by *I Taste Red: The Science of Tasting Wine* (Goode
2016), *Flawless: Understanding Faults in Wine* (Goode 2018),
and now the third edition of *The Science of Wine from Vine
to Glass,* Jamie Goode is a master author of wine books that occupy
the space between popular and technical expositions.

"This is not meant to be a textbook, covering the whole of wine
science in a methodical manner. …I have set out to tell wine science
stories in a way that would engage people who are not overly scien-
tifically literate," (p. 7) he assures us.

While this is generally the case, he cannot help exposing his PhD
in plant biology, along with his command of chemistry, throughout
this important volume. Nevertheless, the acclaimed blogger and wine
writer successfully accommodates the less knowledgeable through his
engaging style while offering insights and opinions that should appeal
to the more informed reader.

The first and second editions of *The Science of Wine* appeared in
2005 and 2014. Regarding the latest edition, "Overall around half
the book is new," (p. 7), Goode tells us. Material covered in his two
books mentioned above is included in abridged form.

Section 1, "In the Vineyard," contains seven chapters covering
the biology of the grapevine, terroir, soils and vines, climate, and
caring for vines.

Section 2, "In the Winery," comprises twelve chapters, including
such topics as microorganisms, flavor chemistry, phenolics, extraction
and maceration, sulfur dioxide, wine faults, *élevage,* sweet wines, and
differences among tasters. Color photographs are sprinkled throughout. A seven-page glossary defines many important terms, occasionally

in more depth than in the main text. An otherwise helpful seven-page index suffers from inconsistent indentation due to entries being listed in four narrow columns, resulting in some confusion as to which subentry is associated with which main entry. Surprisingly, there is neither a bibliography nor a reference section.

In addition to displaying considerable proficiency himself, Goode incorporates quotations and examples gleaned during interviews with or from papers by an array of international experts, most of whom are on the frontlines of research or practice, to illustrate and reinforce points.

The lack of dates left me wondering about the currency of the information, which is particularly important since knowledge is rapidly advancing. The absence of citations of the literature leaves the reader without a way of delving deeper into a topic. In any case, I certainly learned a lot and appreciate Goode's erudition, thoroughness, and readability. I also applaud him for taking positions, even when I don't agree with them.

Chapter 2, "Terroir: how do soils and climate shape wines?" contrasts insights from Australian winemaker Jeffrey Gosset: ("'I don't see winemaking as part of terroir but rather that poor winemaking can interfere with its expression and good winemaking can allow pure expression.'" (p. 27), with anti-terroirist California winemaker Sean Thackery: ("'My objection is simply that [terroir is] so ruthlessly misused... It's very true that fruit grown in different places tastes different. In fact, it's a banality, so why exactly all this excess insistence?'" (p. 29)

Whether one can actually taste the soil in a wine is one of the most fascinating issues tackled. Goode weighs in: "As a scientist who has a working knowledge of plant physiology, I find this notion, which I call the 'literalist' theory of terroir, implausible..." (p.30). This perspective is reinforced by viticulturist Richard Smart and Professor Jean-Claude Davidian of the École Nationale Supérieure d'Agronomique, but fuller consideration of the subject is given in the next chapter.

In Chapter 3, "Soils and Vines," Goode considers the question: "How is it that soils seem to be so important for the wine quality, when science indicates that they are only playing a limited role in influencing the flavor of grapes?" (p. 39) He mentions a 2011 paper, sans citation, by Claire Chenu *et al* on the role of microorganisms, a hot topic that is bringing us closer to what actually gets into the vine and grapes that affect the flavor of the wine. One wonders why more recent work isn't discussed.

Inevitably, the term "minerality" emerges. While calling it "a really useful descriptor," (p. 50), Goode acknowledges that "it's also a term that means different things to different people" (p. 50), begging the question: What does he mean by useful? He quotes a couple of wine writers who claim that the term didn't appear until sometime in the 1980s or later. As I've previously noted (Hulkower 2019), I used "minerally finish" in a tasting note in 1976, a term that I must have picked up from somewhere.

The subsections on "How Experts Use the Term," "Taking Minerality Literally," "Reduction as Minerality," and "The Taste of Terroir" offer additional insights to those that I gained from the work of Alex Maltman (Maltman 2018) and Parr et al (2018), neither of whom are mentioned. In a victory of his right brain over his left, Goode admits: "I used to favor the more established scientific viewpoint, assuming that volatile sulfur compounds could explain much of minerality. But I'm increasingly drawn to the idea that minerals in wine, derived from soil, could be affecting wine flavor in interesting ways…" (p. 54) We'll see.

Chapter 8, "Yeasts and bacteria," contains the best overview of the role of these microbes I've seen. Topics include cultured and spontaneous fermentations, wild yeasts versus cultured yeasts, and seemingly oxymoronic cultured wild yeasts. The table on p. 110 relates classes of compounds produced by yeasts with their impact on flavor. The subsection, "Malolactic Fermentation," is especially good.

Goode largely maintains accessibility for nontechnical readers by defining terms and acronyms along the way and employing his well-honed conversational writing style. He does tend to repeat himself frequently, which, on the surface, might seem unnecessary, but on reflection can be helpful in keeping important points in the forefront.

But then, from Chapter 9, "Wine Flavor Chemistry," that will surely bring nods of recognition from chem-nerds of a feather: "…Marlborough Sauvignon shows quite high levels of methoxypyrazines. These are a group of compounds including 2-methoxy-3-isobutylpyrazine (MIBP; known widely as isobutyl methoxypyrazine [though not by me]), 2-methoxy-3-isopropylopyrazine (MIPP; known as isopropyl methoxypyrazine), and 2-methoxy-3-secbutylpyrazine (MSBP; known as sec-butyl methoxypyrazine)." (p. 134)

Thankfully, these outbursts are few in number and can be scanned or skipped without losing the gist of the discussion.

Goode shines in Chapter 10, "Phenolics," in which he tackles the "fiendishly complicated topic [,] …one where our understanding is incomplete." (p. 140) His explanations of the various chemicals, including tannin and anthocyanins, that are part of the group labeled "phenolics" is essential reading for anyone regularly using those terms.

I work in the tasting room of a small winery in Oregon that specializes in 100-percent whole-cluster-fermented pinot noir and was impressed by a piece by Goode (2012) that is the best I have read on the subject. So, I was pleased to see that Chapter 12, "Whole-cluster and carbonic maceration" incorporates parts of the article while elaborating on the current thinking and practice of this still controversial but increasingly popular approach. After presenting the pros and cons, he concludes: "What was once regarded as an outmoded practice—including stems in red-wine ferments—is now becoming a fashionable winemaking tool for those seeking elegance over power…" (p. 160).

I, too, have noticed that the technique has been increasingly embraced by winemakers in the Willamette Valley over the last decade with delicious results.

Chapter 14, "Wine faults: where are we, and when is a fault a fault?" provides a valuable summary of the material in *Flawless*. Chapter 15, "The evolution of *élevage*: oak, concrete, and clay," is an excellent comparison of the various vessels used to age wine. The table on p. 181, "Flavors from oak," is especially helpful. I was impressed that Goode mentions the Oregon winemaker and creator of terracotta amphora, Andrew Beckham, in the subsection "Clay Around the World," since his work isn't all that well known even in his own state.

Even though a lot of the material covered may be too detailed and nerdy for the novice, *The Science of Wine* isn't suitable for those wanting to master viticulture or enology as a profession. Instead, its value lies in providing a less formal but still in-depth overview of the main areas in each of these two disciplines and serving as an excellent reference. As such, Goode's book belongs on the shelf of everyone involved in any aspect of the wine industry, from producer to writer to consumer.

References

Goode, J. (2012). Stemming the Tide. *The World of Fine Wines, 37*, 90–97. https://worldoffinewine.com/uncategorized/stemming-the-tide-4869650.

Goode, J. (2016). *I Taste Red: The Science of Tasting Wine*. Oakland: University of California Press.

Goode, J. (2018). *Flawless: Understanding Faults in Wine*. Oakland: University of California Press.

Hulkower, N. (2019). Book Review: *Vineyards, Rocks, & Soils: The Wine Lover's Guide to Geology* by Alex Maltman. *Journal of Wine Economics* 14(2), 217-220. doi:10.1017/jwe.2019.19.

Maltman, A. (2018). *Vineyards, Rocks, & Soils: The Wine Lover's Guide to Geology*. New York: Oxford University Press.

Parr, W., Maltman, A., Easton, S. and Ballester, J. (2018). Minerality in Wine: Towards the Reality behind the Myths. *Beverages* 4(4), 77, 19 pages. http://dx.doi.org/10.3390/beverages4040077.

Drinking with the Valkyries: Writings on Wine by Andrew Jefford, Académie du Vin Library Ltd., 2022, 272 pp., ISBN: 978-1913141325 (hardback), \$35.

The foreword to this new book is written by Jay McInerney, whose *The Juice: Vinous Veritas* (2013) I found slight and pretentious. But here he does himself proud, likely inspired by reading the mini-masterpieces he introduces. Spending time with the works of one of the finest, most literate wine writers can do that. Gorgeous writing about noteworthy wines not only increases the pleasure of reading to a higher level but also serves as a model for one's own work.

Drinking with the Valkyries collects pieces originally written by Jefford between 2007 and 2022 and revised in 2021–2022. Most originally appeared in Britain's two "premier grand cru" wine publications, *Decanter* and *The World of Fine Wines*, with a couple in *Noble Rot* and three of the earliest in *Waitrose Food Illustrated*. A brief biography of Jefford and his preface, "Why Wine?" are followed by ten themed chapters, each containing one-to-fourteen articles. An eight-page glossary presents definitions of some wine terms. The Chronology maps each article to its original source by year. The Acknowledgments are followed by an index of eleven pages, each with three columns.

Chapter One, "Origins," contains a single essay, "Homo Imbibens: The Work of Patrick McGovern." It summarizes the investigations of McGovern, who has dedicated his career to discovering the beginnings of viticulture.

"Some Soils, Some Skies," the second chapter, is a fourteen-part tour of the wine world, with stops in both well-known areas like Napa and Bordeaux and little-known wine regions, including Japan and England.

For example, "Downhill All the Way" describes a visit to Mont Granier in Savoie where jacquère, a descendant of gouais, yields wine that "whispers stone rather than singing fruit." (p. 40) In "A Sea Interlude: 2015 Picpoul de Pinet, Cuvée Anniversaire, Beauvignac," Jefford praises the bottle in question: "Picpoul de Pinet is a quiet wine …limpid, sappy, fresh, like unaccompanied flute variations, or classical guitar fandanguillo, coming and going on the wind…" (p. 48). He then follows with an aphorism: "What is truly fine wine is not that which is expensive …but that which is both beautiful and unique …" (p. 48). Descriptions such as these fill the pages and are the pinnacle expression of the "wordoir" of British wine writing.

The thirteen pieces comprising Chapter Three, "Taste and Tasting," offer their own gems. The chapter starts with "Bags, Butter and Biscuits," an alliterative title—Jefford loves alliteration and, backed by a vast colorful vocabulary, deploys many examples—for a critique of wine-tasting language. He insists that "Those who say that all discussion of wine should be in 'plain English' are plain wrong. The results would be plain boring." (p. 67)

He goes on to offer tips on how to communicate about wine. After enumerating features that should be addressed, he advises that you should: "… open your mind to the wider possibilities the wine suggests: not simply flavour analogies but any telling metaphor drawn from your own experience…" (p. 68). Jefford follows his own advice, lacing tasting notes with allusions to music, art, and literature.

"Through the Mangrove Swamp" explores the best way to taste wine. Jefford acknowledges that "No wine is ever fully and satisfactorily assessed until it has passed through the back of the mouth, down the oesophagus, and into the digestive system…" (p. 69). Since professional wine tastings often preclude swallowing, he concludes: "tasting without drinking is a monstrous flaw in all wine criticism…" (p. 69).

Nine distinctive descriptions of "Some Beautiful Wines" comprise the fourth chapter. In "Jewelled Absence: 2016 Petit Chablis, Les Crioux, William Fèvre," Jefford likens its slow ripening to a Dickensian character: "It took all summer to limp its way, like Magwich in irons, to a 12% ripeness…" (p. 99). Of the wine's nose and palate, he writes: "It's the presence that is almost an absence..." (p. 98). With images like this, who needs an aroma wheel?

"The Cup that Consoles" is the sole item in Chapter Five: "A Tea Break." It is a revealing discussion of humankind's second most favorite drink, after water. "Tea is much more widely drunk than wine, not least because of Islam's proscription of alcohol" (p. 122), Jefford explains. A brief history is followed by descriptions of the types of teas and their respective health benefits. Why so much information in an anthology of wine writing? "Both tea and wine ...are old friends of humanity...If I had to choose between the two, I'd choose tea ..." he admits (p. 122).

"Interrogation and Impieties," the sixth chapter, offers thirteen contrarian perspectives on a variety of wine-related topics. *Drinking with the Valkyries* is Jefford's case for not waiting decades to drink vintage port: "...you won't fully understand unless you have tasted it young, in its 'Ride of the Valkyries' stage, when it comes hurtling out of the glass and puts the screamers on you..." (p. 144). "In Praise of Young Wine" expands on this theme: "I love youth in wine, and everything that goes with it: the energy, the excitement, the flesh, the vivacity, the extravagance..." (p. 163).

On a subject about which I've written extensively (see, for example, Hulkower 2019), "The Party's Over" is Jefford's review of the work of geologist and winegrower, Alex Maltman, author of *Vineyards, Rocks, & Soils: The Wine Lover's Guide to Geology* (2018). The party to which he refers and then urges an end to is the persistence of words in tasting notes referring to rocks and other minerals in the literal sense. Maltman, in his book and many articles in popular and academic publications, reminds us that "Any mineral solutes present in wine ...exist at levels well below the threshold for detection..." (p. 158).

Though he regularly deploys "mineral" as a descriptor himself, Jefford has come to see the light: "Maltman's painstakingly argued critique made me aware of the difficulties inherent in drawing any direct inference about aroma and flavor from vineyard soil and geology. I'm happy to use 'mineral,' 'stone,' or 'earth' in a strictly metaphorical sense..."

(p. 159). Given that there is no consensus as to what "minerality" in a wine signifies, an entry in the glossary would have been helpful in understanding what it means to him.

Chapter Seven, "Wine Shadows," focuses on aspects of wine that are disastrous or annoying in a half dozen short essays. Topics include hail, wildfires, earthquakes, price inflation, and "The Curse of the Vertical."

"Score Rigid" is the height of incongruity. "My view is that scores are foolish, philosophically untenable, and damage wine culture rather than enrich it…" (p. 187). So far, so good. But then we get "Readers (and editors) like scores, so of course I use them: refusing to do so would be pompous and unhelpful. We exist to serve readers…" (p. 187). Since I don't regard pandering in contradiction to one's views to be justifiable, I vehemently disagree. For one whose wine writings are the pinnacle of erudition and literacy, Jefford is in a position to subvert this dominant innumerate paradigm and offer a better way to enlighten his audience.

"Wine In A Life," Chapter Eight, collects a dozen vignettes about the joys and travails of a career in wine comingled with descriptions of notable bottles in the context of the circumstances of their consumption. "Mille Fois Morte, Mille Fois Revécue: 2008 Chateau Musar Blanc" starts inauspiciously: "I don't like this wine…" (p. 211). But then, Jefford recalls the late winemaker and manager Serge Hochar who said, "They are difficult, but whenever somebody gets there, they are hooked…" (p. 212). Jefford waits: "…I come back again four days later …and at last, I like the smell of it, fresher than even now, as if dusk has come round to dawn …Serge, you were right…" (p. 213).

A philosophical essay entitled "Wine and Astonishment" is the only item in Chapter Nine, "Against Wine Worldliness." Jefford defines wine worldliness as "a taking for granted of the givens of wine, and the assumption of a kind of assurance or familiarity over the subject that precludes astonishment…" (p. 235), the latter being essential to its fullest enjoyment. Here, again, he criticizes scores: "The scoring of wines is a form of wine-worldliness. It does, of course, acknowledge difference …Yet it also freezes difference by rendering it numerically immutable…" (p. 235). A foolish inconsistency, perhaps, from one who also states, "There's a sort of fun behind scores …" (p. 187).

The title of Chapter Ten, "Three Last Wines," tells it all. It closes the compilation with a trio of wine notes as stories. "Restoration: 2018 Saint-Mont, La Madeleine de Saint-Mont, Producteurs Plaimont" imagines Jefford walking the road to Santiago de Compostela and stopping at a monastery along the way, in the town where the wine originates, for food and shelter: "… this is just what I would want the wine to be: black, sturdy, restorative. Wine that fulfills just a little of the function of blood…" (p. 243).

Jefford does not avoid controversy. On the question of whether or not wine is art, he is clear. In "A Honeycomb of Light: 2010 Mas del Serral, Pepe Raventós" in Chapter Two, he asserts, "… no wine is a work of art" (p. 61). This is echoed in "Wine's Transactional Flaw" in Chapter Seven. Regarding great wine, he insists: "These are a matter of craft, not art…" (p. 176).

And he offers a clear delineation in "Lessons from the Laureate" in Chapter Eight: "A work of art …is wholly a creation of the human mind, whereas the winemaker simply stewards the transformation of one product into another. Winemaking is craft, not art…" (p. 217). His criterion only considers the creator and not the audience.

Yet wine can elicit an aesthetic or emotional reaction as intense as a work of art that, to me, suggests it could legitimately be called one. Also, isn't a sculpture—that results from transforming a natural medium, wood or stone, for example, into another "product"—considered a work of art?

Here's a paradox: Can wine writing, clearly a creation of a human mind, be considered a work of art even though its subject isn't regarded as one by the author? Jefford's work shows that it can. *Drinking with the Valkyries* is one of the finest anthologies of wine writing I have read. McInerney concludes his Foreword by declaring: "It's not wine writing. It's writing…" (p. 13). I'll go a step further. It's superb literature.

References

Hulkower, N. (2019). Book Review: *Vineyards, Rocks, & Soils: The Wine Lover's Guide to Geology* by Alex Maltman. *Journal of Wine Economics* 14(2), 217-220. doi:10.1017/jwe.2019.19.

Maltman, A. (2018). *Vineyards, Rocks, & Soils: The Wine Lover's Guide to Geology*. New York: Oxford University Press.

McInerney, J. (2013). *The Juice: Vinous Veritas*. New York: Vintage Books.

If there is one person I want to write like if I grow up, it's Andrew Jefford. So, it was flattering to get a "thank-you" email from him after he saw this review. It begins: "I see that your extensive review of the Valkyries book has now been published, so this is just a little note to thank you so much for the generous comments you have made about the book. These mean a lot to me."

He goes on to address some of my criticisms: "From an ethical perspective, you are quite right to take me to task for a measure of hypocrisy regarding scores. All I can say in my defence is that I wouldn't have survived as a wine writer without having made that gesture to potential employers, both financially for the family and in terms of 'blind'-tasting exposure. (No one ever pays more, alas, for erudition and literacy in wine writing; some editors regard it as a blemish.) I hope, nonetheless, that the book has a subversive effect here.

"I enjoyed your question as to whether winemaking might be seen as analogous to sculpture in terms of both being a transformation of raw materials, which at best can achieve the status of art. I think there is a difference nonetheless. Given the ubiquity of ambient yeasts, a tank of Lafite juice will end up as wine, whereas a block of Parian marble will not crumble away in the Milos sea breezes to reveal a rough outline of the Venus de Milo; no chemical process helped Alexandros of Antioch on his way. [This makes me wonder whether he could agree that a photograph could ever be considered a work of art.]

"Nonetheless, I fully accept your point about audience response and the kinship between our experience of great art and of great wine."

PART IV

Willamette Valley Happenings

Passport to Pinot:
Something to Walkabout

Think of it as the CliffsNotes version of the International Pinot Noir Celebration (IPNC). Passport to Pinot, known only in 2012 as "Walkabout," is the four-hour abridgment of the two-and-a-half-day wine and food extravaganza held annually since 1987 in the groves of Linfield College in McMinnville, Oregon. The 2013 edition on July 28 featured recent bottlings from wineries in Australia, Austria, California, Canada, France, Germany, New Zealand, and, of course, Oregon. Some seventy pinot noirs, about half from the host state, were accompanied by nibbles from fifteen Portland area restaurants and vendors of artisan victuals.

I sampled just under half of the wines being poured. One impression I had was that many of the 2010 wines from Oregon, while well-balanced and typical of cool-climate pinot, seemed to be entering a closed phase. Notable exceptions included the bright fruity-floral Tendril White Label and the feminine Matzinger Davies.

Most of the 2011s, on the other hand, were still nascent with tannins quick to dominate the palate and finish. For example, Antica Terra "Ceras" promises great complexity, while Brick House Dijonnaise hints at elegant spice and fruit. More approachable, though still age-worthy, was the intriguing Big Table Farm Wirtz Vineyard.

Blue Mountain Vineyard & Cellars in Okanogan Falls, British Columbia, was the sole representative from the Northwest outside of Oregon. Its 2010 Reserve displayed a lovely spice and fruit nose and a soft, graceful floral palate.

California wineries comprised the second-largest delegation. Flowers Vineyard & Winery's 2010 Sea View Ridge Estate Vineyard offered fruit and spice but still needs time. Cakebread Cellars, better known for cabernet sauvignon, shared a 2011 Anderson Valley pinot noir that displayed funk, fruit, and flowers on the nose and finished with silky tannins.

The contingent from Burgundy was the third largest in attendance. Tragically, many suffered losses of up to 100 percent in some vineyards, less than a week earlier, due to ravaging hailstorms that pummeled the Côte de Beaune. Maison Ambrose poured a dark-fruited, nicely balanced 2010 Nuits-Saint Georges Vielles Vignes. Domaine Champy, which will have nothing to harvest this year in its Savigny-lès-Beaune vineyard, dispensed a well-integrated 2010 Beaune 1er Cru Les Champs Pimont. Domaine Cyrot-Buthiau's 2011 Pommard 1er Cru Les Arvelets presented a bright nose of dark fruit, lavender, and spice, with rough tannins signaling the need for more time.

The Southern Hemisphere was represented by Australia and New Zealand. The former's single delegate, Ten Minutes by Tractor, treated tasters to a mature 2009 Estate from high-elevation sites on the Mornington Peninsula. Nautilus Estate from the latter land presented its 2011 Marlborough, which, though immature, had an elegant nose of red fruit and citrus.

But the Passport to Pinot "experience," to use a term that is perhaps too fashionable at the moment, isn't just about the wine. After all, pinot pairs effortlessly with so many foods. A subtitle for the event might even be "Fanfare to Fine Fare." Coquine Supper Club of Portland served up cornbread with pickled pig's tongue, green beans, and basil. Briar Rose Creamery, headquartered in Dundee, offered its complete selection of aged goat cheeses, one of which—Brigid's Bender soaked in Ponzi Vineyard's 2010 Tavola Pinot Noir lees—was particularly appropriate for the event. McMinnville's Walnut City Kitchen ladled an inspired chilled watermelon and hazelnut soup. Levant, another Portland restaurant, served grilled Bharat-spiced lamb leg kebobs on a smoky eggplant relish.

Attendees also got their "just desserts." Ken's Artisan Bakery, in the Nob Hill section of the Rose City, delivered flawless cannelés, a memorable contrast to the forgettable sad, soggy interpretation I had had in Paris in June. Salt & Straw, also in Portland, scooped up Grandma Malek's chunky Almond Brittle with Salted Ganache and surprisingly subtle and delicious Arbequina Olive Oil ice cream.

For the first time ever during intermission at this event, when the wines being poured were changed, a jolt was administered to the revelers, particularly helpful to those who forgot to spit. A bit of Portland weirdness in the guise of the spacey silver-and-white-garbed LoveBomb Go-Go Marching Band invaded the scene. Music most raucous yet undeniably infectious ruptured the growing lethargy, offering a palate-cleanser for the spirit. I simply couldn't stop giggling as skimpily clad females slinked and squirmed to the blasts of brass and the beat of drums. The musicians stopped in front of the assembled winemakers, who were equally enthralled and to whom renewing energy was passed. Brilliant it was and invigorating.

This was my third Passport to Pinot, and I'll likely never tire of attending. While it would be nice to taste more mature wines, practical considerations prevented more than just a few from being offered. The oldest I noticed was from 2008, an Oregon vintage that is still regarded as immature.

It would also be nice to encourage the food vendors to match the wines with more care. While a shrimp salad is quite refreshing and exotic, strong spices enliven many dishes, and these are not the most appropriate for pairing with pinot. A wonderful addition to the menu would be a classic pairing with wild Northwest salmon. Locally foraged mushrooms would also be great.

Near pinot-perfect weather, with temperatures just under eighty degrees enveloped the stately yet comfortable Linfield Oak Grove, nature's gift to the experience. But alas, despite its half-day duration, 2013 Passport to Pinot concluded all too quickly.

Details on future events can be found at http://www.ipnc.org/.

The wines of the year 2010, which were still immature in 2013, blossomed into representations of a truly outstanding vintage. I have never had any Oregon pinot noir from that year that was less than stellar. The year of the typhoon, 2013, on the other hand, turned out to be more of a mixed bag.

IPNC for the Rest of Us

Since 1987, the International Pinot Noir Celebration (IPNC) has regaled lovers of the greatest red wine grape who could endure two-and-a-half days of exquisite food and ethereal drink and afford the ticket now running almost a thousand dollars. For those who have neither the time nor the money, there is Passport to Pinot. The 2014 staging of this four-hour walk-around was held on Sunday, July 27, following the longer event. The Oak Grove of Linfield College in McMinnville, Oregon, provided some shelter from the pinot-hostile ninety-plus degree temperatures.

Attendees sampled featured wines from seventy-two producers from Argentina, Canada, France, Germany, Italy, and New Zealand, and three US states, accompanied by small plates from fifteen local establishments. I was able to taste about a third of what was being poured. With few exceptions, the pinots presented were from 2011 and 2012. Many examples would support declaring either vintage the year of acidity.

From the Northwest, Oregon was represented by thirty producers, Washington State by two, and British Columbia by just one. The still young but promising Brittan Vineyards 2012 "Gestalt Block" exhibited an herbal nose with spice and floral notes that complemented the fruit. The Chehalem 2011 Reserve from the famed Ridgecrest

Vineyard showed great balance with earth and fruit playing nicely together. The delicate, elegant, yet age-worthy Dominio IV 2012 "Mulberry Street" Vitae Springs was redolent of flowers and baking spices. Syncline of Lyle, Washington, dispensed its 2012 Celilo Vineyard Pinot Noir, which had a bright nose with citrus inflected red-candied fruit, high acidity, and moderate tannins.

California producers sharing tastes numbered eighteen. While still tight, the promising Navarro Vineyards 2012 Deep End Blend had a floral and spicy nose and nice structure, with good acidity and tannins. Rhys Vineyards 2012 San Mateo, fermented with 60 percent whole-clusters, produced aromas of earth, coffee, and dark fruit, and an immature finish with lots of acidity and tannins.

France sent one producer from Alsace and nine from Burgundy. Domaine Saint-Rémy from Wettolsheim featured a perfumed 2012 élevé en barriques, with muted floral and earth notes, high acidity, and a delicate palate. As a class, the burgundies I tasted, though quite young, were among the most expansive on the palate. Maison Ambrose teased tasters with an immature 2012 Nuits-Saint-Georges "en Rue de Chaux," which tantalized with understated complex aromas of spices, flowers, and earth. Domaine Charles Audoin 2011 Fixin "Le Rozier" hinted of earth on the nose and balanced fruit, with savory notes on the palate.

Though still young, Domaine Dublère 2011 Volnay "Les Pitures" was soft and floral, with great acidity. Joseph Drouhin offered a delicate floral fruity 2011 Fixin "Clos de la Perrière" that had excellent balance and a long finish. Maison Marchand-Tawse 2011 Gevrey-Chambertin displayed bright fruit and flowers on the nose and lots of earth and acidity on the palate.

Argentina and New Zealand comprised the delegation from the Southern Hemisphere. The 2012 Barda from Bodega Chacra in Northern Patagonia was something of a revelation, with funky aromas competing with floral notes and an elegant palate and with lots of acidity and good tannins. Winery proprietor, Piero Incisa, who showed interest in my connection to the American Association of Wine Economists, promised to stay in touch and could become my latest friend in the industry. Central Otago's Akarua 2010 Akarua revealed bright fruit and flowers and nice acidity. Ready to enjoy now, the serious Villa Maria Estate 2010 Taylors Pass offered dark fruit and great acidity.

The single producer from Italy, J. Hofstätter of Alto Adige, poured a 2011 Barthenau Vigna S. Urbano with black cherries prominent on the nose and lots of acidity.

The culinary offerings were particularly pleasing this year. Red Hills Market in Dundee served an inspired and particularly pinot-friendly wood-fired chanterelle and Briar Rose chèvre crostata. Portland's Boke Bowl wrapped mushroom and tofu "larb" salad with toasted rice and mint salad. Newcomer Ruddick/Wood of Newberg dished up pickled and grilled beets with ricotta and mustard greens. Smoked duck terrine with arugula and pickled radish from rye in Eugene and duck pâté en croute with toasted cocoa nibs from Sybaris Bistro in Albany blended classic elegance with Northwest sensibilities. Coquine Supper Club of Portland plated grilled pork ribs with plum mostarda, fresh white beans, and fennel.

Two Portland establishments provided sweets. Ken's Artisan Bakery returned with meticulously crafted quintessential cannelés. Salt & Straw scooped four highly creative flavors of ice cream, including strawberry honey balsamic with black pepper.

At the midpoint of the walkabout, as wineries cycled out at the event's pouring sites on the periphery to make room for the next round, a procession of winemakers passed through the crowd, waving their national flags. This was followed by the entry of a half-dozen acrobats who provided an amusing diversion during the brief intermission. As many are members of the Confrérie des Chevaliers du Tastevin, the winemakers concluded half-time with "Ban Bourguignon," a traditional song to honor someone or to simply celebrate a good time.

Having now been to four in the past five years, I can confirm that while not as "grand" an experience as IPNC, Passport to Pinot is an easy habit to get into and one I would encourage any "Pinotphile" to consider. It provides a rare venue to sample some of the world's best bottlings with carefully matched foods and to mingle with some of the best producers of our favorite wines. Visit http://www.ipnc.org/ for details on upcoming events.

In Oregon, the two vintages featured, 2011 and 2012, yielded quite different wines. The first was unusually cool, with grapes ripening slowly and brought in at the beginning of November ruining Thanksgiving for winemakers. On the other hand, 2012 was a dry summer, with the first rain coming on October 12, the only day I volunteered to help pick pinot noir for Dukes Family Vineyards. After more than a decade, the wines from both vintages have evolved as expected. The 2011s are leaner, less about fruit, and more savory, with ample acidity. The zaftig 2012s highlight fruit and are richer. Each produced outstanding examples of Oregon pinot noir.

The heat experienced during the event foreshadowed 2014 emerging as the first of five hot vintages in Oregon.

Piero Incisa and I never connected.

A Random Walk through IPNC

When gorgeousness is bottled, the label reads pinot noir. A particularly apposite gathering place for those who share this belief was the twenty-ninth edition of the International Pinot Noir Celebration, held from July 24 to 26, 2015, on the Linfield College campus in McMinnville, Oregon, during a very brief respite from record-breaking heat.

Master of Ceremonies Sam Neill—an actor and proprietor of Two Paddocks, a Central Otago, New Zealand, winery—welcomed attendees to the Church of Pinot Noir, offering several incentives to embracing membership like acceptance of sin. He then asked rhetorically, "Have I taken this obsession with a grape too far?" As one who has adjusted his life and selected my residence based on its proximity to major producers, I'm certainly not one to judge. It was, however, a particular delight to be an invited media attendee able to savor the entire event and share two-and-a-half days with my co-religionists.

With nearly three decades of practice, the organizers settled on a schedule that kept the faithful well-fed and well-moistened, from seven-thirty in the morning until late into the night, both Friday and Saturday. The remarkable breakfast buffet featuring Oregon bounty was followed by a vineyard tour and winery lunch one of the days, and The Grand Seminar, lunch on the lawn, and a University of Pinot seminar on campus on the other.

Afternoon activities included white-wine tastings, food samplings, and book signings. Each afternoon, half of the sixty-eight featured wineries from Australia, Canada, France, Italy, New Zealand, South Africa, and the United States poured the celebrated beverage at alfresco tastings. The Grand Dinner, a sit-down affair, completed the first day. The legendary Northwest Salmon Bake was the penultimate event on Saturday night, with the Sparkling Brunch Finale mercifully scheduled for ten a.m. on Sunday morning.

This schedule merely served as a framework since one's experience at IPNC was largely shaped by random encounters with other attendees and the beverages sampled. After opening ceremonies, I boarded a bus to a winery, the identity of which was revealed only when we arrived. Stunningly sited Colene Clemens Vineyards hosted us for a tour, a seminar on clones, and tastings of their pinot noirs as well as those from Bien Nacido Estate and Davis Bynum in California and Coelho Winery and WildAire in Oregon. Guest Chef Katy Millard of Coquine, which opened just five days earlier in Portland, prepared appetizers and a hedonistic lunch, accompanied by wines from our host.

Wine & Spirits Magazine senior correspondent Patrick Comiskey moderated the Grand Seminar, entitled "Tasting the Stars: Champagne & Sparkling Wine." Producers and distributors from France, California, and Oregon shared details of the challenges faced in making this most celebratory drink, as well as samples of their bottlings. The theme was chosen in honor of the fiftieth anniversary of the first planting in the Willamette Valley of pinot noir by David Lett. His son, Jason, shared a unique sparkling wine made by his father.

I attended the University of Pinot seminar, "Does Vine Age Matter?" led by Burghound Allen Meadows. We tasted Oregon pinot noirs produced by Beaux Frère, Bethel Heights, Elk Cove, and Ponzi, from both old and young vines. The overwhelming sense was that old vines produced more interesting wines.

The casual setting for the afternoon activities facilitated self-directed encounters with various vendors. Elephants offered tastes of three cheeses produced by small dairies around the country and write-ups to answer the question: "Why is this cheese so damn expensive?" Nicky USA provided John Gorham of Portland's Toro Bravo and John Sundstrom of Lark in Seattle with raw materials to create recipes from their cookbooks.

Particularly striking was a comparison of an Oregon pinot noir sampled in both a Riedel Oregon Pinot Noir glass and a burgundy bowl. The nose from the latter was relatively flat and muted while the aromatics from the former would bring smiles to the true believers. Floral whites, viognier, riesling, and gewürztraminer, produced by some of the featured wineries, were poured the first afternoon while chardonnay joined the celebration the second.

Two hours each afternoon were allocated to the alfresco tasting, during which I was able to taste about thirty of the featured wines. One difficulty is that many are still immature and, given the circumstances, none could be aired adequately to compensate.

Noteworthy entries from outside the host state were the spicy 2012 Eldrige Estate from Australia, a citrusy 2012 Hanzell Vineyards from California, a delicious 2012 Aloxe-Corton "Clos de la Boulotte" from Domaine Nudant in Burgundy, and a young but intriguing 2012 Barthenau Vigna S. Urbano from J. Hofstätter in Italy.

Thirty-four featured wineries were from the Beaver State. Among the standouts were an elegant 2012 Antikythera from Antica Terra, a warmly fruity 2012 Estate Reserve from Belle Pente Vineyard & Winery, the refreshing Big Table Farm 2013 Willamette Valley, a nicely balanced Crowley 2012 La Colina, and a tight but complex Kelley Fox 2012 Maresh Vineyard.

While I met many Oregon wine industry celebrities—including Maria Ponzi, Luisa Ponzi, Tony Soter, Brian O'Donnell, and Rollin Soles, whose contributions are now the stuff of history—my more memorable and sustained chance encounters were with those who were either relative newcomers or from other pinot noir-producing regions.

On the bus ride to and from the winery, I sat next to Christina Pollan, an elementary school teacher from La Mesa, California, who, along with her brother, established Pollan Family Vineyards in 2014 in the Yamhill-Carlton District. She hadn't been into wine until three years ago when her sibling suggested she start tasting if she was going to be marketing their grapes. Since their first "vinifiable" harvest isn't expected until 2017, she is just beginning to think about potential buyers.

At the Grand Dinner, I was seated between Lorna Kreutz, recently named winemaker at featured winery, Lincourt Vineyards in Solvang, California, and her assistant, Ryan Aura. She poured us two examples of small-production pinots from 2012, Willie Mae and Annie Dyer. The former was aged in 40 percent new French oak and showed mint and pepper on the nose. It was nicely balanced and delicate. Floral and spice notes characterized the Annie Dyer, which saw 60 percent new French oak. These were among the California pinots that are improving my opinion of pinots from that state, Oregon snob that I am.

At the lunch on campus, I sat between Ms. Pollan and featured winery Lemelson Vineyards' newly appointed winemaker, Matt Wengal, who moved to Oregon from California to replace Anthony King, now at Carlton Winemakers Studio. Lemelson has tended to use a higher percentage of new oak, and Matt plans to bring that down a bit.

He poured two library wines for us. The 2000 Jerome Reserve saw 57 percent new oak. After some airing, it was elegant, with some mint apparent. Counterintuitively, the 2001 Jerome Reserve, which was aged in 87 percent new oak, showed less influence from the barrels than did the 2000. After breathing a bit, we enjoyed the classical cool vintage pinot with lean fruit and mint.

At the Salmon Bake, I sat with Marina and Bill Knutson, proprietors of SpierHead Winery, a tiny producer in Canada's Okanogan Valley, with whom I had also lunched at Colene Clemens. I was the only non-Canadian at the dinner table, where I met the proprietors of one of the featured wineries, the eponymous Meyer Family Vineyards, also in the Okanogan. I had no problems communicating with them, however, since I had learned to speak Canadian, albeit with an Eastern American accent, when I worked in Toronto during the summer of 1969.

Pinots from each of the two were nicely structured and well-balanced. Should global warming push us further north in search of traditional examples, it's nice to know that some promising work is already being done there. Our sommelier ensured that we were poured samples of a seemingly endless array of mostly pinots from around the world, one dating back to the 1980s. I also flagged down Brian O'Donnell to get a sample of his 2002 Belle Pente Reserve from a magnum; not surprisingly, it was a gem.

The Sparkling Brunch Finale was a bit of a free-for-all, with buffet tables covered with an eclectic array of eats and surrounded by attendees eager to savor the last moments of the event. Costumed women showered us with giant bubbles as we consumed several examples of the featured beverage, along with an even more insane selection of foods than had been served the previous two days.

Having had the chance to amble about the friendly and unconstrained confines of a very well-planned event, I can assure my fellow fanatics that all is well in Mondo Pinot. If you're interested in joining the annual celebration next year, visit www.ipnc.org.

I am told that the 18.8 acre Pollan Family Vineyard appears on the latest map of the AVA. I have no additional information about where the grapes are being sold. Also, WildAire Cellars has closed. On the other hand, both Canadian wineries mentioned are still active with SpierHead, named for Spiers Road where it is located in Kelowna, British Columbia, now written SpearHead. Brian O'Donnell continues to produce lovely, reasonably priced wines that are capable of aging gracefully.

My in-depth coverage of the seminar can be found at https:// www.researchgate.net/publication/282107439_Pinotgogy_Celebrating_and_contemplating_the_grape.

As this book was coming together, it was announced that 2024 would be the last IPNC in its current form. Demand was dwindling and costs were rising. Ticket sales were sluggish. However, I have it from a reliable source that after some reorganization, it will be back.

My reports about the Oregon Chardonnay Symposium, later the Oregon Chardonnay Celebration, from 2015 to 2017, were never published, so I offer them here. These events occurred at the beginning of the revival of the grape's reputation in Oregon, which had been unfairly maligned. Happily, Oregon chardonnay is now highly regarded and represents an excellent, albeit distinctive, alternative to increasingly pricier white burgundy.

O Chardonnay

Whether chardonnay is the best of grapes, the worst of grapes, or something in between is on the palate of the taster. On March 14, 2015, the fourth annual Oregon Chardonnay Symposium, hosted by Stoller Family Estate, pressed the case for Willamette Valley expressions of this white Burgundian cultivar to be recognized on par with its pinot noir. The four-hour event, themed "Attack of the Clones," was equally divided between a Technical Session and a Grand Tasting. Stoller's stunning tasting room was extended and tented to provide shelter from the late winter rain.

Technical Session

Rajat Parr, a partner and proprietor of Seven Springs Estate in the Eola-Amity Hills, as well as of two other wineries in California, moderated the panel discussion. Addressing history, Jason Lett of The Eyrie Vineyards led off with an attack on the clones, encouraging us "to stop talking about clones, ironically, by talking about clones." He asked, since they don't talk about clones in Burgundy, why should we talk about them here?

He also provided evidence that challenged the impression that the early clones planted in Oregon were lackluster. His 2012 Original Vines Reserve from the heritage Draper Selection was extremely elegant, with honeyed perfume, lovely texture, and a fine mid-palate. The balance was excellent, with nice acidity and a bit of toast on the finish.

John Paul of Cameron Winery took a scientific/philosophical approach. He differentiated between clones and vineyard selection and pointed out that these are frequently confused. His 2012 Clos Electrique Blanc from heritage clones offered an intense honeyed aroma, with flavors after the nose. It was more bold than elegant and finished long, with good acidity.

Craig Williams, owner of X Novo Vineyard in Eola-Amity, spoke about diversity. He advocated clonal diversity for complexity and the health of the vineyard. The 2013 Walter Scott X Novo Vineyard Chardonnay was a clear reflection of this disjointed vintage. The nose was very toasty with floral and spice attempting to push through. The flavors were somewhat diluted, but the finish was reasonable.

Thomas Bachelder of Bachelder Wines compared chardonnay vineyards in Oregon, Niagara, and Burgundy. The latter are mostly field blends of various clones. Oregon and Niagara plant wider than higher. The aroma of his 2011 Johnson Vineyard bottling initially burst forth with very ripe tropical fruit but then gave way to chalk. Some juicy fruit was noted on the palate, with a hint of toast and a nice mid-palate. It opened up with air.

Mimi Casteel of Bethel Heights Vineyard talked about how to farm. She stated that while the differences between clones aren't interesting, they are important. Because there is no skin contact, it's hard to relate terroir to chardonnay.

Nevertheless, since it will grow anywhere, where the cultivar is planted must be selected carefully to get something elegant and transparent. The aroma of the 2013 Bethel Heights Casteel Reserve led with wet stone and juicy fruit, but while elegant on the palate, it was light and short, again reflective of the vintage.

Rajar Parr's presentation addressed his 2007 Seven Springs Estate "Summum" Chardonnay. Hazelnuts dominated the bouquet, with light toast and some fruit. The entry was delicate and demur while the mid-palate was bolder. The finish was long and the overall impression was classy, elegant, and confident.

As John Paul aptly summarized: "Clones absolutely matter, and the more the merrier."

Grand Tasting

Only fifty-six of the more than one hundred chardonnays sampled by the organizers were selected to showcase. I tasted sixteen during the two-hour session. Here are some highlights.

The 2012 Antica Terra Aurate displayed toasted hazelnuts, great balance, nice acidity, and a long, yummy finish. A 2012 Beaux Frères Gran Moraine offered a similar experience. Bergström poured both the 2012 and 2013 Sigrid and, as they were representative of each vintage, they couldn't be more different. The 2012 was big, buttery, and rich while the 2013 was much more restrained and untypically bright. A 2010 Belle Pente Belle Pente Vineyard exhibited toasted hazelnuts over fruit, an elegant mouthfeel, and medium-long finish. One of the more complex pours was the 2013 Big Table Farm, which showed spice as part of the flavor mix. Another delight was a mouth-watering 2013 Brittan Vineyards Chardonnay with a very floral nose.

I was particularly impressed with bottles from the warmer 2012 vintage, which reined in the fruit and struck a balance with the other components. These included the typically restrained 2012 Brick House Cascadia and the nicely balanced 2012 Matzinger Davies. Two particularly food-friendly wines were the 2012 Soter North Valley Reserve, with its laser-focused aroma and bright citrus flavor, and the

2012 Stoller Family Estate Reserve Chardonnay, with an amalgam of aromas and rich flavors.

Final Thought

My nose and palate lead me to prefer the more restrained style of chardonnay with crisper fruit, spice, chalk, or wet stone, and in older examples, a touch of honey. A long, lingering finish and sufficient acidity to accompany a range of foods are necessary as well. While the most memorable chardonnays I have enjoyed are burgundies, it is clear that several Oregon producers are well on their way to matching their success with pinot noir and fielding compelling New World offerings. O Chardonnay, you are certainly nothing to duck.

While the name I gave to the piece is taken from a Cajun song, it also sets up the pun in the last sentence.

Fêting Chardonnay
the Oregon Way

"To drink a series of white burgundies, or Chardonnays, as you might several clarets at a sitting, with the object of comparing them, is unusual," observes Hugh Johnson in *A Life Uncorked.* On February 27, 2016, two hundred and fifty wine industry members and consumers gathered for the fifth Oregon Chardonnay Celebration (OCC) at The Allison Inn & Spa in Newberg, Oregon, to do just that.

Highlighting the growing importance of Burgundy's premier white variety in the New World's best location for its premier red variety, the International Pinot Noir Celebration (IPNC) is now a partner in the OCC. This alliance was first announced at an afternoon tasting of chardonnays produced by some of the invited winemakers at the 2015 IPNC. A name change from Oregon Chardonnay Symposium and a move to posher quarters signaled more ambitious plans.

The Media Dinner

The evening before OCC, a handful of media invitees, including this writer, gathered at Adelsheim Vineyard for a preview of the main event. At the reception, winemakers from Adelsheim, Alexana, Chehalem,

Evening Land, Ponzi, and Stoller, and the managing director of Domaine Drouhin Oregon poured samples of 2013 and 2014 chardonnay and discussed their approaches to dealing with each vintage.

We were then invited into the handsome dining room to enjoy a three-course dinner prepared by the winery's chef, accompanied by one new and several older vintages. Seared diver scallops were matched with Domaine Drouhin Oregon's floral, rich 2014 Roserock Chardonnay from a recently acquired vineyard in the Eola-Amity Hills American Viticultural Area (AVA) and Alexana's complex 2012 Willamette Valley Chardonnay, its first wine from that grape.

Next came spaghetti carbonara with pancetta and parmesan, accompanied by Stoller's crisp, youthful 2011 Reserve Chardonnay and Adelsheim's juicy, food-friendly 2010 Caitlin's Reserve Chardonnay. The third course, chicken scallopine, was matched with a trinity of older vintages, which evoked solemn admiration. Chehalem's 2008 Ian's Reserve Chardonnay and 2007 Seven Springs Estate "Summum" Chardonnay were both youthful, with great acid balance. The polished Ponzi 2006 Chardonnay Reserve was riper and lusher.

Each convincingly demonstrated that Oregon chardonnay can comfortably age along with the best of them regardless of annual variations. Winemaker Luisa Ponzi opined: "Oregon Chardonnays miss some of the drama of Oregon vintages since they are picked early."

Panel Discussion and Tasting

A case study: Interpreting Chardonnay filled the first two hours of the OCC. Arrayed in front of us were five glasses containing samples of wine made by five different winemakers from Dijon clone 96 chardonnay harvested from the Lark Block of the Durant Vineyard in 2014. Two were component wines destined to be blended, two were barrel samples of what will be single-vineyard bottlings, and one was a complete wine already bottled and released.

Leading off the session, moderator Elaine Chukan Brown observed that chardonnay is the most site-expressive and most expressive of the winemaker's intent. Willamette Valley chardonnays display a constellation of tiny flavors that accumulate to give the same amplitude as those from California. This Sonoma County resident proclaimed that Oregon chardonnays are unique, delicious, and fine.

Paul Durant detailed the characteristics of the Lark Block of his eponymous vineyard. A wheat field until the late 1980s, this 200-foot elevation sedimentary soil site was planted with Dijon clones of chardonnay in 1991.

Brian Marcy of Big Table Farm offered one of the component wines, which showed bright floral notes with a flash of anise. His goal is simply delicious and balanced wine. He picks when he can obtain complete fruit requiring no adjustment. The wine fermented from twenty-two brix to bone dry in four-and-a-half weeks and underwent malolactic fermentation.

Thomas Bachelder of Bachelder, Oregon, contributed one of the samples of what will be a single-vineyard bottling. It had a juicy nose of concentrated fruit, which continued on the palate with good acidity, nice balance, and a long finish. The wine went through malolactic fermentation. Bachelder pointed out that to get good acidity from grapes from a vigorous site like the Lark Block and avoid fruit bombs, one needs a minimum amount of fruit. The more time the wine spends in the barrel, the more varietal features fade, and "you can see the face of the stone."

In other words, you can taste the land. He also avoids talking about angel's share, the wine that evaporates from barrels but, rather, says to look down instead of up as a cook does when reducing a sauce.

Joe Dobbes of Dobbes Family Estate showed a component wine, which displayed bright lemon, richer flavors, and lower acidity. It will be blended with equal amounts of Dijon clone 76 from the same vineyard. He used 30 percent new oak and no malolactic fermentation. He picks his barrels based on flavor, noting that bigger isn't better, and looks for richness, mouthfeel, higher acidity, texture, and primary and tertiary fruit. Chemistry is a secondary consideration. He loves Burgundy and brings back yeast from there.

Marcus Goodfellow of Goodfellow Family Cellars brought a wine that will be bottled as a single-vineyard chardonnay in another five months. Its nose and palate were dominated by toasty oak with a burst of juicy fruit flavor. He advocates slow fermentation, followed by slow *élevage* (barrel-aging) in larger barrels to prevent the wine from becoming tired.

As is done for white burgundy, his stylistic model, he uses a mixture of barrel types including barriques, puncheons, and larger casks. The wines only undergo malolactic fermentation in one of these. His goal is to have the wine taste like the site and have intense flavor. He picks early to get brighter acidity.

Paul Durant is a farmer, not a winemaker, so his contribution, the sole finished offering, was made by Isabelle Dutartre. It exhibited medium intensity, nicely balanced fruit, rich flavors, a bit of heat, and a medium finish. It did complete malolactic fermentation.

Brown noted the savory element and fruit core in all of the samples. While Brian Marcy said that winemakers are still trying to find their voices, Brown, as an outsider, thought it was a really exciting time for Oregon chardonnay, with more media attention, increased investment, and plantings. She urged us to remember this: Oregon chardonnay is on the verge of exciting changes.

Grand Tasting

For the next two hours, the emphasis turned from technical and educational to festive. Chardonnays from forty-four wineries in the Willamette Valley and one each from Southern Oregon and the Columbia Gorge AVA were poured, in most cases by their producers. I tasted twenty-one, many of which needed to be warmed or allowed to breathe a bit before sampling. Three from the 2015 event were poured again and, once again, merit mentioning: the elegant Beaux Frère 2012 Gran Moraine Chardonnay, the yummy Brittan Vineyard 2013 Chardonnay, and the very pretty EIEIO 2013 Cuvee O Chardonnay.

Other standouts included the polished Crowley Wines 2013 Four Winds Chardonnay, the floral Longplay Wine 2013 Lia's Vineyard

"Jory Slope" Chardonnay, the hunger-inducing Matzinger Davies 2013 Chehalem Mountain Chardonnay, the herbal REX HILL 2013 Seven Soils Chardonnay, and the chalky ROCO Winery 2014 Chardonnay.

Concluding Thoughts

Like the wine it regales, the OCC is an event that is rapidly maturing into an important annual vinous affair. The first years were focused on technical aspects and aimed at producers who were at the vanguard of promoting chardonnay as a major variety, as logical as pinot noir is for our state. And while discussions among wine industry peers remain central to the event, consumers are now invited to participate as well.

The result is that the seminar walked the line between delving into the technical aspects of producing distinct Oregon chardonnays and addressing some of the most basic questions from nonexperts regarding winemaking and wine tasting. That can be a lot to ask, given the time limitations.

According to the Full Glass Research report dated January 2015, chardonnay plantings in Oregon in 2013 covered 1,164 acres, a distant third behind pinot noir and pinot gris. And though that same year, 2,605 tons of chardonnay were brought in, almost up to the 2,846 tons in 2000, Oregon still has a long way to go before this variety attains the pre-eminence that its red burgundian counterpart has. That the IPNC has embraced this event is an important step toward the marketing aspect of gaining this grape its rightful place. As we leave the Anything But Chardonnay (ABC) error behind, we should now be asking Why Not Chardonnay? We'll know it has arrived when the O in OCC is changed to I for "international."

The North, South, East, and West of Oregon Chardonnay

The Seminar

The seminar that launched the 2017 Oregon Chardonnay Celebration at The Allison in Newberg on February 25 took us on a tour. Ray Isle, the executive wine editor of *Food & Wine*, served as moderator and guide as we went "Roadtripping through Oregon Chardonnay Country."

"People overlook how incredibly complex the wines can be," Isle asserted. He quoted winemaker Anna Matzinger, who observed that a "high amount of intellectual capital is being applied to chardonnay in Oregon." Five distinct examples from around the state substantiated her claim.

Bob Morus, of Phelps Creek Vineyards near Hood River, represented the East. His 2014 "Lynette" Chardonnay had a pretty floral and fruity aroma and was sleek on the palate, with nice acidity and salinity.

Luisa Ponzi, of Ponzi Vineyards in the northern part of the Willamette Valley, advised that when deciding to pick, "you want to catch [chardonnay] right before it tastes great." Her 2014 Aurora Chardonnay had a toasty, fruity, and nutty nose and a rich, balanced flavor.

Maggie Harrison of Antica Terra gets her chardonnay from Shea Vineyard on the Western side of the Willamette Valley. She admitted that she doesn't "know how to pick before it tastes good." Instead, she picks part when the acid is good, then lets the clusters sit to let the flavors develop. The 2014 Aurata Chardonnay offered complex aromas of oak and fruit in an elegant dance. Similarly, the beautiful palate was rich and evolving, with a long, layered finish.

Heading a bit south, we heard from Ken Pahlow of Walter Scott. The 2015 X Novo Chardonnay is from a young vineyard in the Eola-Amity Hills, planted to fifteen clones. The nose was dominated by toast and nuts but with air, spice and fruit emerged. The balance and acidity were nice, but there was little fruit on the palate. It was clear that it needed more time.

The tour ended in Southern Oregon. Bryan Wilson of DANCIN Vineyards discussed his 2015 "Melange" Chardonnay from grapes grown at an average of 1,800 feet in elevation. Initially muted, the wine sat zaftig on the palate, with some juicy fruit and richness but little acidity. Again, this will benefit from additional aging.

Isle summarized the tour by highlighting the focus, tension, and acidity common among the chardonnays featured. While "elegant" is a term that can be fraught since it might mean thin and lacking power to some, these wines offered both elegance and power.

The Grand Tasting

We then adjourned to the Grand Tasting. For two-and-a-half hours, forty-six wineries, including the five featured at the seminar, poured their chardonnays from either the 2014 or 2015 vintage. Appetizers including mushroom popovers, deviled eggs, and smoked steelhead trout prepared by Jory lent savor to balance the acidity and complement the richness of the wines.

In addition to the five served at the seminar, I sampled twenty-eight bottlings. In general and not surprisingly, the 2014s were less acidic and more fruity. In contrast, the 2015s were better balanced and immature but showed great promise for ageability.

From the older vintage, the standouts were the mouth-filling offering from Brittan Vineyards; the lemony but lingering Crowley

Wines "Four Winds;" Chehalem's complex Ian's Reserve; and the sleek Grochau Cellars' Bunker Hill Vineyard. The Matzinger Davies and the Evenstad Reserve from Domaine Serene were two that were particularly food-friendly.

Promising chardonnays from the younger vintage included the attractive Knudsen, the yummy Fairsing, the bright Big Table Farm, and the juicy Willamette Valley Vineyards Bernau Block.

Reflections

It seems that chardonnay never actually fell completely out of favor despite the now-faded "Anything But…" movement. It remains among the most planted grape varieties in the world, second among whites to Spain's airén. Naturally, as with pinot noir, Oregon wine-growers have looked east to Burgundy rather than south to California in search of models of chardonnay greatness.

What we now are seeing is the beginning of the payoff of the "intellectual capital" that is being expended up and down the state. While there will never be a single style of chardonnay in Oregon, just as there isn't in the Côte-d'Or, it is more distinguished and dis-tinguishable from what comes from California. No liquid-buttered popcorn or oak splinters here. Instead, balance and acidity are king.

This structure suggests greater ageability, most recently for the vintage 2015 bottlings. The good news is that more producers around the state are embracing the grape, even grafting over the less profitable pinot gris to it. If things keep up as they have been, Oregon chardonnay will be ready for a Cole Porter-style tribute:

I love the smell of you, the lure of you

The fruit of you, the pure of you

The nose, the legs, the mouth of you

The east, west, north, and the south of you

I'd love to gain complete control of you

And handle even the heart and soul of you

So love, at least, a small percent of me, do

For I love all of you

This report did appear in an edited version in the Oregon Wine Press *under the title "Coming of Age."*

Time's Way with Oregon Chardonnay

The Seminar

The eighth edition of the Oregon Chardonnay Celebration at The Allison Inn & Spa in Newberg on February 23, 2019, attracted two hundred seventy wine industry members, consumers, and media interested in sampling current and older releases of a white wine variety enjoying resurgent interest and acclaim. The seminar that opened the event, entitled "Time in a Bottle: The Evolution of Chardonnay," was moderated by Jason Lett of The Eyrie Vineyards and featured a panel comprised of well-respected longtime Willamette Valley chardonnay growers and producers.

Robert Brittan of Brittan Vineyards, Ben Casteel of Bethel Heights, Michael Etzel of Beaux Frères, and Kate Payne-Brown of Stoller Family Estate each provided two chardonnays. One was from the cool 2010 vintage, in which the grapes struggled to ripen. It was remembered for a late harvest that provided food for migrating birds to bulk up for the long flight south at the expense of yields.

Care was taken by the sommeliers when pouring the potentially fragile wines: the glasses were gassed with argon and covered. The other wines were from a much warmer vintage, either 2015 or 2016, both of which ripened rapidly and were brought in earlier than normal, in some cases almost two months before the 2010 vintage.

In his introduction, Lett noted that the celebration was being held on "the day that we enter the fiftieth year of growing chardonnay in the Willamette Valley." He told us that early on, his father, David "Papa Pinot" Lett, along with others, actually planted more chardonnay than the grape he is known for. Over the years, the percentage of planted acres of chardonnay declined from 38 percent in 1981 to 17 percent in 1998 and to 6 percent in 2017. While the latest figure might seem pitiful, it's on the rise. In 2014, the number was under 5 percent and has been increasing since then. Still, chardonnay ranks a distant third in planted acres in Oregon behind pinot noir and pinot gris.

Lett also stated that Oregon chardonnay was a critical success early on, with noted writers commenting on the similarities to its counterparts in France. He also proclaimed, "Chardonnay is one of the great aging varieties," a point reinforced by what we were tasting. He emphasized that the site, not chemistry, makes chardonnay age-worthy.

Robert Brittan, who credited chardonnay as one of the reasons he came to Oregon from California, shared the 2010 and 2015 Brittan Vineyard Chardonnays. The former had a youthful nose, with a broad flavor palate and long finish. A later visit revealed candied fruit and a hint of anise on the nose. The younger wine was medium-gold, a bit darker than the 2010, and emitted complex aromas of nuts, fruit, and spice. The flavors were nicely balanced, delicate, and youthful, and the finish was long.

Ben Casteel brought the 2010 and 2016 Bethel Heights Vineyard Estate Chardonnays. The older wine, which was the only 2010 presented that was bottled under a screw cap, was a warmer yellow-green, with a mature nose and broad fruit flavors that seemed to be fading. In contrast, the 2016 was yellow, with hints of green, had nutty aromas, initially, with an unusual woodlike scent later, and was youthful on the palate.

Michael Etzel offered his 2010 Beaux Frères Ribbon Ridge and 2016 Yamhill-Carlton Chardonnays. With a bit of funk on the lemony nutty nose, the older wine was really delicious. The classy 2016, silvery-yellow with a hint of green, had a polished fruity nose and palate. In the past, Etzel harvested at higher brix but now brings the grapes in at around twenty brix.

Kate Payne-Brown contributed the 2010 and 2015 Stoller Family Estate Reserve Chardonnays. The 2010 was showing its age but was still drinking well. I enjoyed the butterscotch on the palate that emerged with some air. The younger example had a lovely bright nose and round flavors. Payne-Brown underscored Stoller's approach of exploiting its warmer site to get more fruit but balancing it with good acidity.

The panelists shared their histories of growing chardonnay and views on producing a wine from it. They compared picking decisions, percent of new oak, use of concrete and amphora for fermentation, oxygen management, and closures. The consensus was, as Payne-Brown advised, "Do farm for ageability," and, as Brittan recommended, practice "winemaking with intent." He also acknowledged: "Chardonnay has been the greatest teacher…of all the wines I've worked with over the years…"

The Grand Tasting

Fifty featured wineries, including the five mentioned above, shared samples of their chardonnays over two-and-a-half hours, accompanied by hors d'oeuvres. I tasted twenty-six from twenty-five producers. Many were still young, which augured a promising future. Not all the bottles were at optimal temperature nor had enough time to breathe.

Nevertheless, thirteen stood out: the scintillating 2017 00 VGW Chardonnay, the distinctive 2016 Bergström Sigrid Chardonnay, a balanced 2016 Domaine Divio Chardonnay, the lovely 2016 Domaine Drouhin Oregon Roserock Chardonnay, the tasty 2015 Dukes Pearl Chardonnay, a promising 2016 EIEIO Yates Conwill Vineyard Chardonnay, an intriguing 2016 Elk Cove Goodrich Chardonnay, a toasty 2016 Flâneur La Belle Promenade Chardonnay, a really good 2015 Johan Vineyards Visdom Chardonnay, the bright 2016 Lavinea Elton Vineyard Chardonnay, a beautiful 2017 Lingua Franca Estate Chardonnay, a nice 2017 VIDON Estate Chardonnay, and the complex 2016 Winderlea Chardonnay.

Reflections

Ageability is a hallmark of most great wines. The panelists agreed that the 2010 chardonnays would go on for at least another five years. I noted my concerns about Bethel Heights, though there may be some bottle variation at play. All were drinking well at over eight years old, and some will certainly go a few years longer. My vote of confidence that time will continue to be kind to Oregon chardonnays is the stash of 2013, 2015, 2016, and 2017 bottles I have laid down for my grandchildren. Time will tell if we can keep our hands off until the kids are allowed to taste them.

Below is the lightly edited urtext of the heavily edited version of the 2023 article that can be found at https://www.guildsomm. com/public_content/features/articles/b/neal-hulkower/posts/ willamette-valley-chardonnay.

Willamette Valley's Way with Chardonnay

I haven't heard much about the Anything But Chardonnay (ABC) rebellion lately. Good. One purpose of this article is to add momentum to the Try Oregon Chardonnay (TOC) movement that has been growing for at least a decade. While the grape can be found in a few other regions in Oregon, 83 percent of the planted acres of chardonnay are in the Willamette Valley based on 2022 figures. As Josh Bergström of his eponymous winery suggests, "I would use the term 'Willamette Valley Chardonnay' over 'Oregon Chardonnay' as most of the qualitative and stylistic increases have come out of the valley and not from the southern or eastern parts of the state."

One characteristic of a wine considered fine is its capacity to age. As I discovered, many of the excellent older Willamette Valley char- donnays I sampled certainly confirm that. Some remained youthfully fresh while others grew wiser, deeper, and more complex. Only a few were on their last legs.

Ian Burch, winemaker at Archery Summit and a panelist at the Oregon Chardonnay Celebration Seminar in February 2023, concluded: "I think that Willamette Valley chardonnay can age very long and very well." Jason Lett, proprietor and winemaker at The Eyrie Vineyards notes: "You will find that old Oregon chardonnay drinks extremely well. The pre-mox epidemic that struck white Burgundy [beginning] in the nineties never came to Oregon. If you love older white Burgundy, then you'll find many incredible bargains by picking up old Oregon chardonnay at auctions and cellar sales."

Lett's observation alludes to the second purpose of this article, which is an exploration of how Willamette Valley chardonnay performs in the market vis a vis white burgundy. My curiosity was piqued by a remark Chevonne Ball of Dirty Radish made during the 2023 Oregon Chardonnay Celebration Seminar that she moderated: "The price and the getting of French chardonnay is so high that it is less expensive for people in the UK and in Asia to get Oregon chardonnay...so people are buying it because it's less expensive but also because it's delicious."

My investigation, which included visits to some prominent wine retailers in London and communications with importers and producers in the Willamette Valley, revealed a far more nuanced and complex situation.

Willamette Valley Becomes Chardonnay

David Lett planted the first chardonnay vines in the Willamette Valley in 1965 and produced the first wine from them in 1970. It turned out to be a brilliant decision. Situated around the 45th parallel, the Willamette Valley, like Burgundy which is closer to the 47th parallel, is in a sweet spot for growing chardonnay. (On the other hand, Bordeaux, which also sits around the 45th parallel, has only a tiny amount of the grape planted.)

Andreas Wetzel of his eponymous winery agrees: "I feel that Chardonnay, grown in a cooler climate such as the Willamette Valley, allows us to craft wines that retain brighter finishes. They are refreshing, help accentuate the fruit/floral notes, and are a pleasure to serve with foods that are full-flavored. Furthermore, when keeping in mind Chardonnay's aging potential, a well-balanced chemistry benefits from the cool winds that we see in the Van Duzer Corridor. This translates to wines that cellar well and grow more complex in nature."

Though pinot gris remains the most planted white grape in the valley, chardonnay has become the one to watch. It is finally emerging from undeserved obscurity and attracting new consumers. Plantings are increasing and even some pinot gris vines are being grafted over to it. There is something of a bandwagon effect going on.

Should We Care about Clones?

This question can still be a trigger for religious skirmishes among producers. Eyrie grows three clones, primarily a Draper massal selection, along with Wente and Sterling. Other early producers in the valley also planted clones from California with mixed results.

"There is a pervasive narrative that 'Oregon chardonnay was bad but now it's good.' That is objectively false," insists Jason Lett. He notes that some notable wine critics recognized early-on its excellence but, at the same time, others denigrated it.

Consequently, the overall reputation of Willamette Valley chardonnay suffered despite the undeniable successes. Many thought that California clones were inappropriate for the cooler, wetter climate, so the so-called Dijon clones were brought in from Burgundy starting in 1984 and have been popular ever since.

Although early unhappiness with Oregon chardonnay had more to do with site selection, farming, winemaker competence, and overoaking than with the use of California clones, I still heard some strong opinions claiming Dijon clones are behind the resurgence of Willamette Valley chardonnay.

On the other hand, Simon Davies, buying director of A&B Vintners, the largest importer of Willamette Valley wines in the UK, insists that "Clones are boring." He believes that the site is the important thing.

While many winegrowers stick strictly to planting a handful of Dijon clones, others favor a much wider selection, including some from California, since climate change has greatly reduced the risk of grapes not ripening. For example, Abbott Claim is establishing a vineyard with fourteen different clones of chardonnay in the Eola-Amity Hills, which is cooled every afternoon by breezes through the Van Duzer Corridor, promising to produce wines of greater complexity and freshness as well as long ageability.

My position on clones leans toward that of Simon Davies. The wines that I sampled that are aging beautifully were made anywhere from a single clone to several. Some were only Dijon clones while others were a mixture of Dijon and California. What seemed to matter most was the vintage and the winemaking and not the clone.

Capacity to Age

When asked if anything extraordinary is done to ensure that chardonnay will age, Véronique Boss-Drouhin, the head winemaker at Domaine Drouhin Oregon, quipped, "I would love to tell you that I do something extraordinary like dance around the barrels once a month on a full moon night! I believe the aging potential is already in the fruit, but there certainly are some wise decisions that probably do contribute to help the wine age."

Bergström agrees that aging potential is derived both from the grapes and the choices of the winemaker. He explains, "The regional climate of the Willamette Valley ensures ageability in our wines due to the high natural acidity and the fact that we achieve physiological ripeness/good flavors at the same time that natural acids are high and potential alcohol is low. But, ageability with white wines, especially chardonnay, [requires] making sure that oxidation does not happen prematurely, so our wines are made in a reductive winemaking style."

Jay McDonald, owner and winemaker at EIEIO & Company, takes the opposite approach. He picks at a higher level of titratable acidity and pre-oxidizes the juice before fermentation, which occurs at colder temperatures.

Shelby Perkins of Perkins Harter Wines has tried both approaches: "Primarily, pick date matters…To me, natural acidity is the essence of the structure in a white wine and should never be adjusted. So, pick decisions are critical. Also, I began pressing grapes in a protective manner, with oxygen excluded, but have moved toward the brown (or 'black') style of must treatment for ageability." This style exposes the must to hyperoxygenation, which turns it dark and counterintuitively protects the resulting wine from oxidizing early, thus extending its life.

My research confirmed that there are many paths to success.

Tasting Older Willamette Valley Chardonnays

Let's get the obvious question out of the way first: What do I mean by old? I arbitrarily picked vintage 2017 and older, or at least five years old. As you'll read, this turned out to be too recent. With that criterion, I tasted a few dozen examples from vintages 1991 through 2017, during the spring and summer of 2023, which ranged in age from five to thirty-one years old. The results were a revelation and changed my opinion of when to start drinking chardonnay.

The Oldest: Still Alive but Past Their Prime

The Eyrie Vineyards has been producing chardonnay in the Willamette Valley since 1970. Jason Lett has some from the first vintage, but it's not for sale. He teases that "the 1973 is still stellar." He poured the 1991 Reserve, a vintage ideal for aging, which came from "The Eyrie" vineyard. The nose was juicy-fruity, tilting to very ripe tropical fruit, with hints of nuts and maderization. While past its peak, it was still drinking okay, with a rich but shorter finish and low acidity.

The 1997 King Estate Reserve Chardonnay was deep-golden, with a hint of brown, and had some life left. The nose showed butterscotch with dried fruit. The palate didn't have much acid in the medium finish. On the other hand, the 1996 was definitely in decline.

Doug Tunnell, founder and winemaker at Brick House Winery, served a deep-golden 2002 Chardonnay that was definitely nearing the end but was still good, with a slightly maderized nose showing some vanilla and a rich palate of deeply roasted nuts and madeira.

Jesse Lange, winegrower at Lange Estate Winery & Vineyards, brought out the 2003 Reserve Chardonnay and a 2004. The older, sealed with a plastic cork, was maderized. Though in decline, a 2004 Lange Estate Freedom Hill Chardonnay had a creamy fruit nose and an elegantly rich pretty palate, with some juicy fruit, good acidity, and medium-short finish.

A 2009 Domaine Drouhin Oregon Arthur Chardonnay had a deep-lemon color and a waning fruity nose of ripe mango and hints of licorice. Arron Bell, assistant winemaker and operations manager, pointed out that the "intention is not to make age-worthy chardonnay" but to strive for freshness.

A 2012 Iris Vineyards Chardonnay, while past its prime, was still drinking reasonably well. The nose showed olive and with air, oak. The palate had hints of oak, vanilla, and pineapple, and a short finish.

Older but Still Fresh

What surprised and delighted me was that a few older examples were still youthful. Not unexpectedly, cooler vintages are doing best.

Ian Burch, winemaker at Archery Summit and formerly with Evening Land Vineyards, poured chardonnays from both, but only those from the latter were old. The 2007 Evening Land Seven Springs Vineyard "Summum" from the inaugural vintage, was bright and fresh, perfumed, saline, and mouthwatering. The 2010 Evening Land Seven Springs Vineyard "La Source" offered an intense lemon-lime and mango nose, rounded texture, and great acidity.

A 2011 EIEIO Yates-Conwill Vineyard Chardonnay was still youthful, with green notes and light toast on the nose. An impressively textured, lovely, delicate palate resolved with an elegant, rounded, mouthwatering finish.

Mature Yet Still Going Strong

More frequently, those between seven and twenty years old showed richness and complexity not found in younger wines and, in several cases, still had many years of life left.

Belle Pente Vineyard & Winery proprietor and winemaker Brian O'Donnell shared a 2003 and a 2008 chardonnay. The older one, the first made from 100-percent estate fruit, was deep-gold with a nose that opened up to overripe fruit. The palate was after the nose. While the finish was short, the wine was still going. In contrast, the 2008 had a medium-gold color with silvery sparkle, a toasty nutty nose, and a beautifully balanced mouthwatering palate in which tropical fruit played a supporting role.

From his library, Josh Bergström poured 2008, 2012, and 2013 Sigrid from magnums. All three were very much alive, with the oldest emitting rich tropical fruit aromas and filling the mouth with evolving flavors, an impression of tannins, and a long finish.

Tunnell also shared 2010 and 2011 Cascadia Chardonnays, both from cool vintages. Deep-brass with hints of green, the 2010 had an odd bouillon-like nose but a lovely mouthwatering savory palate. The 2011 was deep-brass with a toasted brioche nose, some juicy fruit, and a nice entry on the palate.

Carabella Wine owner and winemaker, Mike Hallock, and I tasted his 2007, 2008, 2011, and 2017 vintages. The 2007 was the most intense of the older vintages while the 2011 had a lovely butterscotch nose, with ripe tropical fruit as a back note and a crisp, saline palate with some citrus, a medium finish, and ample acidity.

At Chehalem Winery, the house that Harry Peterson-Nedry built, I sampled 2008, 2011, 2012, and 2015 Ian's Reserve and 2017 Reserve Chardonnay. The 2008 offered lovely aromas of light toast and fruit that became more forthright with air. The faintly fruity, nutty palate lingered with a hint of butterscotch and great acidity on the finish.

Domaine Drouhin Oregon also poured the 2012 Arthur Chardonnay. It was "kaleidolfactic," with hints of butterscotch and toasted nuts on the nose and an evolving fresh palate, with a rich, saline, medium-long finish and years to go.

From Goodfellow Family Cellars, Marcus Goodfellow brought out a 2010 Matello Richard's Cuvée. Still doing fine without a hint of tiredness, it showed ripe tropical fruit on the nose. The richly textured palate was nicely balanced with softer acids.

With many years left, the 2006 King Estate Signature Collection Oregon Chardonnay was a medium-deep yellow with some green and faint gold. The nose included iodine, mango, and herbs. Elusive flavors played on the nicely balanced, suave palate.

At ROCO Winery, I tasted with cofounder and winemaker, Rollin Soles. From his previous vinous venture, he poured a 2007 Argyle Nuthouse Lone Star Vineyard Chardonnay followed by a 2012, 2013, 2014, and 2015 made at ROCO. The 2007 still had flashes of green and warm aromas of citrus and bread, with faint pear on the palate and a long finish. Deep-yellow moving toward gold with hints of green, the 2012 from Rose Rock West Vineyard had a bouquet of tropical fruit, white flowers, and hints of nuts and oak. The palate was delicate and nicely textured with a medium finish and a flash of lemon. With air, the nose became intensely lemony, with clean laundry aromas and a bit of toast.

2013: The One Vintage That Consistently Impressed Me

In addition to Bergström's Sigrid, I was particularly fond of several other chardonnays from the 2013 vintage. The aroma of the 2013 Brittan Vineyards initially exhibited chalk and line-dried laundry, with butterscotch pushing up. Later, vanilla wafer emerged. The palate was richly textured with a long finish. Lovely stuff. Wise, with years to go.

The 2013 Fairsing Vineyard Chardonnay had a rich honeyed nose with ripe tropical fruit and a rich toasty nutty flavor, a fresh medium finish, and great balance.

A 2013 EIEIO Yates-Carlton Chardonnay had a rounded fruity bouquet with fresh laundry overtones and a bit of toast. The spherical palate with excellent acidity did a cute flip-up in the finish with an acid burst.

The 2013 ROCO Willamette Valley Chardonnay exuded an intriguing complex aroma of tropical fruit, anise, and line-dried laundry. The well-balanced palate was creamy with great texture and a long finish.

2014 to 2016: Hitting Their Stride

Not surprisingly, the chardonnays from this string of hot vintages are generally maturing sooner.

The beautifully textured 2014 Goodfellow Willamette Valley Chardonnay showed lots of oak on the nose and had a well-balanced palate with excellent length.

Exuding a big citrusy nose that later went to very ripe tropical fruit, the 2014 Fairsing Chardonnay showed a bit of honey on the otherwise savory palate. In contrast, the 2015 was nicely perfumed with tropical fruit, some floral notes, and lower acidity.

ROCO's 2014 and 2015 chardonnays were also drinking well with years to go.

The nose of the 2015 Chehalem Ian's Reserve Chardonnay initially had a touch of reduction, then toast, line-dried laundry, chalk, hints of lemon and butterscotch, and graham cracker. The palate had great texture, excellent balance, and a long finish.

The youthful 2015 Dominio IV Imagination Series Chardonnay was surprisingly crisp for a hot vintage and showed lime-inflected brioche on the nose and a pleasing palate.

Wetzel shared 2015, 2016, and 2017 chardonnays. All were drinking well, but the standout was the 2016, with aromas of nuts, lemon, a hint of fresh laundry, and a complex palate with nice acidity.

The 2016 Perkins Harter Johan Vineyards Chardonnay from a magnum had aromas of toasted nuts, fresh citrus, and with air, butterscotch. The palate was soft with a medium-long finish.

Nearing its peak, the 2016 Goodfellow Willamette Valley Chardonnay had a butterscotch and oak nose and a nicely balanced palate.

Wayne Bailey, proprietor and winemaker at Youngberg Hill Inn & Winery, had but one old chardonnay, a 2016 Aspen. The nose showed tropical fruit over oak and some toast that carried to the rich, well-balanced palate. The acid was good and the finish very long.

2017: Mostly Still Too Young

Most 2017 chardonnays I sampled were still youthful and can't be considered from an older vintage at this writing. The 2017 Eyrie Vineyards Original Vines Chardonnay was shy on its "kaleidolfactic" nose and delicate on the palate, with a long finish.

A 2017 Domaine Divio Willamette Valley Chardonnay had a polished oak nose with hazelnut and a hint of lime. Still very young, the elegant, gorgeously textured palate was understated with a medium finish.

The 2017 Chehalem Reserve Chardonnay seemed muted on the palate but had a lovely seductive nose.

The 2017 Dominio IV Imagination Series Chardonnay had an intensely floral nose with more oak. The lovely mouthwatering palate was a bit tight and crisp, with a hint of juicy fruit and a medium-polished finish.

A 2017 Goodfellow Family Cellars Richard's Cuvée Chardonnay from Whistling Ridge had a rich nutty palate and light toast nose, with some bright stone fruit poking through, but was still young. The same was true of a 2017 Lange Freedom Hill Vineyard Chardonnay.

An immature 2017 Perkins Harter Wines Johan Vineyard Chardonnay poured from a magnum had an intense citrus but tight nose and palate, with a medium-short finish. In contrast, the 2017 La Belle Promenade, also from a magnum, had a richer nose with tropical fruit and hints of citrus, funk, and wood polish. Its palate was young, with a lovely texture and medium finish.

From Stoller Family Estate, I tasted three chardonnays from the 2017 vintage: 100-percent stainless steel-aged Dundee Hills, the Reserve, and Elsie's. The Reserve's aroma developed from faint lemon-lime, floral, and a hint of spice into a more complex array.

The rounded palate echoed the lemon-lime and was a textural delight with a medium-long finish and great balance. Elsie's was classy, with a richer nose of slightly creamy lemon, clean line-dried laundry, and wood. The entry was elegant and lemony with a hint of lime. The finish was medium with nice balance and texture.

Bottom Line

As a result of tasting through an extensive assortment of aged Willamette Valley chardonnays, I concluded that I shouldn't consume any less than five years old. Frequently, I was poured a current release at the end of a flight of older vintages. Invariably, it presented itself as underdeveloped, shy, and immature next to its older siblings. I hadn't realized what I was missing pulling corks on these youngsters. The sweet spot for the chardonnays I sampled was between seven and twenty years old, albeit with plenty of variation depending on the vintage, closure, and winemaking intent.

Willamette Valley Chardonnay versus White Burgundy in the Market: Producers' Perspective

I asked several producers of Willamette Valley chardonnay: If you sell your chardonnay outside the winery in the US, how well does it compete with white burgundies? If you sell your chardonnay outside the US, how well does it compete with white burgundies? Predictably, the opinions and experiences of producers of Willamette chardonnay are all over the map.

Lett was characteristically blunt: "Who cares? This is Oregon wine, and [I] hope it expresses itself that way."

Others wouldn't mind taking on white burgundy in the market but are finding it challenging. Soles lamented, "Unfortunately, international hasn't really taken up the 'better than Burgundy' or 'Burg's replacement' yet. I'm not sure we see our chard as a competitor with Burgundy (after all, Burgundy is HUGE acreage and majority-planted to chardonnay, which means a very wide range of price points). It would be beneficial to our style of chardonnay if retailers and restauranteurs positioned our chardonnay next to listings of white burgundies."

Aaron Lieberman, winemaker at Iris Vineyards admitted: "We still encounter a lot of market resistance to Oregon Chardonnay. In the Western US, I think that resistance is because most chardonnay drinkers have a taste for California chardonnay. In the Eastern US, Oregon chardonnay might be more accepted as more chardonnay drinkers there have a taste for white burgundy." Iris does not sell outside the US.

Then, there are some that have been finding success in the market.

Josh Bergström noted that "our Old Stones Chardonnay was designed to compete with village level and premier cru white Burgundies, especially in markets like the UK, Denmark, and Scandinavia. That is working very well; however, the export market is still very price sensitive, and most importers will shy away from reserve bottlings of chardonnay from the US due to price."

Bruno Corneaux, proprietor and winemaker at Domaine Divio, only sells domestically. "It competes well, in my opinion, as I am from Burgundy and have had many true white burgundies that I feel my chardonnays can stand up to."

Tunnell acknowledged: "There is no doubt that, especially for our London importer, Oregon chardonnay is very attractive to long-time white burgundy collectors." He added, "In both the US and Asia, we have a price advantage."

McDonald said that his chardonnay "competes quite well in the Japanese and Korean markets and did very well in Hong Kong too until they changed who controlled that country."

Bailey stated that his wines successfully compete against white burgundies in the US and Canada. Hallock confirmed, "We've had some success in Ontario/Quebec markets, but their marketing is a little odd up there."

Burch gave a more detailed response: "In general, I think our wine is competing well—although ...it's still not well known that Willamette Valley chardonnay could be a substitute for white burgundies.... Our chardonnays are a good deal versus any premier cru burgundy. The ripeness levels and underlying acidity are not dissimilar.

The wines in the Willamette are getting made more and more like burgundies with many Burgundians playing in the Willamette now…. Both areas embody moderate new oak levels, tightening winemaking techniques… With recent vintages being shorter in Burgundy resulting in higher prices of their wines, this has opened the door for Willamette Valley chardonnays to consumers/buyers seeking a similar style… Our other strong point is the change in wine styles sought by many consumers/buyers for lower alcohol/higher acidity/more aromatic and fresher wines."

Perkins offered an anecdote: "I get a lot of notes from people indicating that they were fooled by my wines blind or that they seem like Burgs or Chablis. A friend who is a seller in NYC recently put my 2021 Chard up against a 2020 Dauvissat a client purchased to drink with him and reported that it held its own and that the client is a massive Burgophile and was more interested in my wine than the Dauvissat."

Boss-Drouhin stated: "We sell our chardonnays at the winery and all across the US, as well as in many different export markets. People who want white burgundy will buy white burgundy, but we see fast growing interest in Oregon chardonnay. I don't think of it as competition, just good choices."

Willamette Valley Chardonnay versus White Burgundy in the Market: Retailers' Perspective

Simon Davies of A&B Vintners finds that white burgundy buyers are a natural audience for Willamette Valley chardonnays. Buyers are equally interested in Willamette Valley chardonnay and white burgundy, with 90 percent buying both. He believes that the biggest difference between the two regions isn't the prevalence of limestone in Burgundy but the diurnal shift.

Climatologist Dr. Greg Jones is researching the difference between Burgundy and Willamette Valley climates. The former is more continental while the latter is more maritime. He notes: "The reason that the Willamette Valley has lower nighttime temperatures is because of humidity levels, which are much higher in Burgundy."

So, Oregon gets cold at night during the growing season, but Burgundy doesn't. This results in the preservation of acid in Willamette Valley chardonnays.

With regard to pricing, Simon Davies noted that some Oregon chardonnays clearly compete with some village and premier cru white burgundies and represent an extraordinary value compared to them. On the other hand, a white burgundy selling for, say £80, is worth £160 on the secondary market. This isn't the case for Oregon chardonnay.

Barbara Drew MW, content officer at Berry Bros. & Rudd, also sees Oregon chardonnay making inroads into the UK market "as part of a wider trend of collectors looking to diversify their cellars. This has been accelerated, at least in part, by small volumes of white Burgundy, such as in 2021."

But like Simon Davies's clientele, it isn't a case of Oregon displacing Burgundy. "[It] is rare to see collectors who are already fans of white Burgundy entirely shifting away from France. Instead, they are seeking out fine, New World chardonnays, such as those from Willamette Valley, to complement their existing collections," she asserted.

Drew called out "beautifully aged examples from producers such as Domaine Drouhin Oregon and Lingua Franca now available in the UK market."

As far as relative value, she found that "Given the price point, these wines sit comfortably alongside premier cru burgundies."

During my visit to London in June 2023, I dropped in on three retailers. At Fortnum & Mason, I saw a 2020 Walter Scott Chardonnay for £29.95, a 2018 Gran Moraine Chardonnay for £60.90, and a 2017 Evening Land Seven Springs "Summum" Chardonnay for £120 in a separate section for North American whites. I was told that white burgundy customers buy white burgundy and not Oregon chardonnay. The latter is stocked for visitors who want a taste of home.

Harrods carried a 2019 Cristom Chardonnay for £40, a 2018 00 VGW Chardonnay for £105, and a 2018 00 EGW Chardonnay for

£130. The store's buyers decide what to carry, so the selections can vary, but I didn't detect any enthusiasm for Oregon offerings.

Hedonism Wines also carried the same Gran Moraine and Evening Land chardonnays as Fortnum & Mason for the same price. I was told that 95 percent of the chardonnay sold is white burgundy and the rest from elsewhere. An Oregon chardonnay occasionally is slipped into a blind tasting of white burgundies to generate interest.

While A&B Vintners is having great success in marketing Willamette Valley chardonnay to white burgundy drinkers, the retail outlets I visited showed little enthusiasm for it and certainly didn't appear to give it any special attention.

If the examples presented above don't convince you to join the TOC campaign, whether as a substitute for the increasingly inaccessible and expensive white burgundies or as a complement, perhaps Boss-Drouhin who has a presence in each region will: "I think the most important thing is that the quality of Oregon chardonnay is incredible, and the wines are delicious and compelling. That's what matters."

Bibliography

Alberty, Michael, (2023). "The Near-Death and Resurrection of Willamette Valley Chardonnay." *Wine Enthusiast* online at https://www.winemag.com/2023/04/18/willamette-valley-chardonnay/.

Cole, Katherine, (2017). "Meet Black Chardonnay." *Seven Fifty Daily* online at https://daily.sevenfifty.com/meet-black-chardonnay/.

Danko, Pete, (2022). "Oregon wine gains ground in UK as alternative to pricey Burgundy." *Portland Business Journal* online at https://www.bizjournals.com/portland/news/2022/03/28/oregon-wine-in-uk.html.

Gaffney, William "Rusty," (2016). "Oregon Chardonnay Gaining Prominence." *Pinot File* Vol. 10, Issue 33, https://www.princeofpinot.com/article/1821/.

Helm, Jade, (2015). "End of the Clones." *Oregon Wine Press*, May 2015, online at: https://www.oregonwinepress.com/article?articleTitle=end-of-the-clones--1430253663--2084--.

Hulkower, Neal D., (2019). "Coming of Age." *Oregon Wine Press*, April 2019, p. 10 and online at https://www.oregonwinepress.com/coming-of-age-chardonnay-celebration-2019.

Institute for Policy Research and Engagement (2023). *2022 Oregon Vineyard and Winery Report* https://industry.oregonwine.org/resources/reports-studies/2022-oregon-vineyard-and-winery-report/#:~:text=Total%20 wine%20grape%20production%20in,2020%20and%2010%25%20 from%202019.

Robinson, Jancis, (2022). "Oregon fills the gap." https://www.jancisrobinson. com/articles/oregon-fills-gap.

The editor of the Oregon Wine Press *asked me to cover this event. Even though I only tried weed a couple of times in the last century and have no desire to do so again, I replied: "As a child of the sixties who is, in fact, free on Saturday, how can I refuse?"*

I include this story not only to give a glimpse of the earliest attempt in Oregon to harmonize two distinct mood-altering substances but also to put back in the word "not" that I, unfortunately, left out in what I originally submitted, thus dramatically misrepresenting my opinion as to whether this is possible.

The Day of Wine and Flowers

Do Pinot and Pot Pair Well?

In his garage office, Marcus Horne and a roommate were enjoying some recreational marijuana while brainstorming ideas for a new business. As it turned out, one was right under their noses. Why not organize wine and weed tours? Only four states—Alaska, Colorado, Oregon, and Washington—have legalized the use of cannabis for purely recreational purposes. Both Colorado and Washington, which passed the measures before Oregon, already have wine, food, and weed tours. But Horne's Homegrown Weed & Wine Tours has him swimming in a blue ocean, with nary a competitor in Oregon in sight at the moment. (I did learn that some wine tour operators will make unscheduled stops to purchase marijuana, but none advertise the fact or allow smoking in the vehicle.)

On June 11, 2016, I met up with six female tourists and three staff at the Western Oregon Dispensary (WOD) in Newberg, the first stop of Homegrown's prelaunch tour. While I was waiting for them to arrive, proprietress Sheri Ralston checked me in and gave me a swag bag. Horne said that when he first approached Ralston with the idea, she immediately wanted to be part of it. Thus heartened, he decided to give it a go.

Comfortably seated in the reception area outside the showroom and sales area, I chatted with that day's guest grower. Juan Lopez of Infinity Farms, whose Gorilla Glue #4 is one of the offerings at WOD, grows his plants hydroponically. The Cuban-born Lopez came to the United States in 1980 with his parents and settled in Louisiana doing electrical work. He moved to Oregon almost two years ago to indulge his passion for gardening. While Lopez does not smoke himself, he enjoys the challenge of growing his product and using his experience to do his own maintenance.

After the others arrived and checked in, we were admitted through the locked doors to the sales area. Brightly lit showcases contained laced edibles and a variety of paraphernalia, which the well-scrubbed sales staff cheerfully described. Being of both the numeric and acronymic persuasions, I had to inquire about two pairs of letters and numbers that were displayed everywhere. For each of the over two dozen cannabis types and on every product that contained marijuana, percentages of tetrahydrocannabinol (THC) and cannabidiol (CBD) were shown on small cards.

The young gentlemen behind the counter explained that THC was the psychoactive ingredient, more important to potheads, while CBD, which interacts with THC, is of interest to those who use marijuana for medical purposes. Of the flower offerings at WOD, THC ranges in potency from just over 7 percent to almost 27 percent, and CBD goes from 0 percent to 11.63 percent.

At this point, it is important to state that I do not smoke or consume marijuana in any form; no buds were burned by me either during the research for or preparation of this article. Nevertheless, I was curious to see and sniff examples of what was available. The attractive glass canisters contained nothing like I saw in the last century. Three types—Quantum Kush (THC 21.59 percent, CBD 0.53 percent), Jazz (THC 25.49 percent, CBD 0.05 percent), and Optimus Prime (THC 26.84 percent, CBD 0.08 percent)—were really quite lovely. The first was pretty with small crystals and a complex pleasant aroma. Jazz smelled grassier, and Optimus Prime was even stronger.

Because of the nature of the product, one can't do weed tastings as one might do a comparative wine tasting. So, I was amused to find a Marijuana Tasting Journal for sale. For each strain, one enters the price, THC, CBD, method, flavors, effects, and notes, and colors in one-to-five leaves to indicate a rating.

Those in search of reefer gladness can select 1.3 gram pre-rolls of Lemon Alien Dawg, Mr. Rodgers, Platinum Girl Scout Cookies, White Widow, Big Wreck, and Obama Kush from a Joint Bar. Edibles include Jolly Greens, F&D Hot Cocoa, and Delta 9 Caramel Corn.

We next embarked on a small bus that seated fourteen passengers on the periphery. The driver, Horne, was required to be walled off from the back to keep secondhand smoke out. On the short drive to our next stop, Duck Pond Cellars in Dundee, a small pipe containing one of the purchases was lit and passed around. When I expressed concern that smoking and getting high would interfere with wine tasting, several attested to the opposite; flavors were enhanced, especially after consuming weed in food. To avoid ill effects, the consensus seemed to be not to partake of cannabis after wine tasting, especially if one has overmoistened. One should remain stone-cold sober if planning to get stoned.

When we arrived at the winery, Paul Johnson hosted the Small Lot Tasting Flight featuring a 2015 pinot noir blanc, a regular and a rosé pinot gris, both from 2015, a 2014 gamay noir, a 2014 pinot noir, a barrel sample of pinot noir from 2015, a 2012 syrah, and a

2006 semillon dessert wine. Interleafing the pours, he took us on an extensive and leisurely tour of the winery. Time flew by so quickly that Duck Pond became our final stop.

For Horne to have a thriving business, the two main ingredients, weed and wine, should be compatible at some level. Both do contain terpenes, which give each floral notes and other pleasing aromas. Whereas the sole purpose of enjoying cannabis is intoxication, this is generally regarded as an undesirable side effect of wine tasting.

Furthermore, smoking and sipping do not generally go together. Vaporizing (or vaping) the buds can possibly work to avoid conflict. I suppose that one could develop vinous pairings of snacks consumed to satisfy the munchies. Looking in another direction, for those in need of a string of wine descriptors and tired of all the flora and fauna foisted on us by the professional "vintelligensia," cannabis in its endless varieties might be just the place to look.

With Homegrown Tours, Horne has expanded the meaning of "roach coach." If you are wondering "Can a bus transport me to even greater heights?" I refer you to Horne and his crew who are most congenial hosts. You will certainly get "bhang" for your buck and the experience could suit you to a tee.

As of the end of 2023, twenty-four states have legalized recreational marijuana.

Homegrown Weed & Wine Tours' website is no longer active. I left the last paragraph in excising only the defunct URL because I thought you might enjoy the puns. I have found no evidence that anything has taken its place. I also could find no evidence that Infinity Farms is still operating, let alone growing cannabis.

On the other hand, Western Oregon Dispensary continues to thrive with five locations and more being planned. The website offers Block Berry, with a THC of 36 percent. Who said you couldn't get much higher?

Battle Creek Flows into the Pearl

With the opening of its first-ever tasting room at 820 NW 13th Avenue, just north of NW Johnson Street in Portland's Pearl District, Battle Creek Cellars has branched out from its home in the Willamette Valley. A preview of the striking 2,000-square-foot facility was held on October 30, 2019, for media and other invited guests. It opened to the public on November 2.

The handsomely appointed multi-use area features a bar with limited seating and indoor and outdoor tables. Without leaving town, visitors can enjoy daily tastings and an assortment of light food options including the obligatory charcuterie.

The first of three estate vineyards, Battle Creek Vineyard was planted in 1998 in Turner by one of the future founders of Seattle-based Precept Wine. Since its founding in 2003, Precept has owned the namesake winery. Battle Creek was the site of the first conflict between tribes and settlers in the Willamette Valley. Yamhela Vineyard, planted in 2007 in Yamhill, and Roe Vineyard, planted in 2008 in Newberg, add diversity of soil types, elevation, orientation, and clones to the winery's sources of pinot noir and chardonnay. Some fruit, including other varieties like riesling, is purchased.

Since 2005, the wines have been made at the 12th & Maple Winery in Dundee. In 2014, Susan Cabot became the winemaker.

She trained at South Seattle Community College's Northwest Wine Academy and went on to work as assistant winemaker at Belle Pente and later as principal winemaker at Omero Cellars. She also served as assistant winemaker at WillaKenzie Estate.

Battle Creek produces 13,000 cases annually, of which 10,000 are the value-priced Unconditional Tier, which has a rosé of pinot noir, a red blend, and a pinot noir. Other bottlings include a Reserve Tier, with a rosé of pinot noir, a chardonnay, a white blend, and a pinot noir, all from Willamette Valley grapes; and a Single Vineyard Tier, comprised of pinot noirs made from each of the three estate vineyards and a Méthode Champenoise Blanc de Noirs made from Yamhela Vineyard pinot noir at Gruet in New Mexico, another Precept-owned winery known for its sparkling wine. The nicely balanced 2014 vintage estate wines were among those being poured. The Amphora Series Tier, a tasting room exclusive, features the results of Cabot's experimental fermentations in a 500-liter amphora from Italy, including a Carbonic Red Blend and a riesling.

Cabot, a Portland resident, looks forward to spending time in the tasting room, not only mingling with visitors and acting as a most knowledgeable brand ambassador but also doing blending trials on site. She explains: "This space is going to be more than just a wine-tasting room…it's going to be a place where people can unwind after work, meet up with friends before dinner, and maybe even watch a big game on the projector screen. Pinot noir is our foundation, but we'll have some really different limited-run wines that are exclusively available at the tasting room for people to experience."

Battle Creek Cellars's shiny urban "wineshed" is a smart new bauble in the Pearl.

Happily, the tasting room in downtown Portland still thrives. Visit https://battlecreekcellars.com/ *for the latest hours and information about the wines.*

Linfield Raises the Bar

Who makes up the most important segment of the wine industry? The wine buyer. While vineyard stewards, wine-growers, and winemakers are essential to creating the product, it is the revenue derived from wine purchases that fuels the business. And the most profitable route is direct-to-consumer sales. Giving a wine lover the ability to try before buying increases the chance of a felicitous outcome.

Recognizing this, the Department of Wine Studies at Linfield University in McMinnville has reincarnated the space that previously housed Elena's at 546 NE 3rd Street, on the corner with NE Ford Street in McMinnville, as a wine bar and learning lab called the Acorn to Oak Wine Experience. Following a soft opening around Labor Day 2023, the bar had its hard opening on September 8.

The idea emerged from focus groups of students in the Wine Studies program who were interested in getting out of the classroom and gaining real-world experience. Planning for this ambitious undertaking began in 2022 under the leadership of Tim Matz, former director of the Linfield Center for Wine Education and Domaine Serene Chair in Wine Business. While working on her master's degree in wine business leadership, Stephanie Mitchell served as project manager for the venture. Now a professor with the WSET Program at Linfield, she handed the reins over to Paul Johnson in August 2023. He became the direct-to-consumer sales manager responsible for day-to-day operations, assembling a team of Linfield students from a range of backgrounds to work in the tasting room.

In 2015, Paul Johnson moved to Oregon from Connecticut, where he had gained experience working in both the wholesale and retail sides of the wine business, to immerse himself in "the most exciting wineries in the New World." He got involved in winemaking at Duck Pond and later opened the Alit Wine Room in Dundee. This experience proved useful for launching Acorn to Oak. He also started his own winery, Satyr Fire, and serves as its winemaker. Most recently, he spent a year at Domaine Roy & fils.

Johnson's team, all Linfield undergraduates over twenty-one, includes Victor Sandoval, a senior communications major from San Marino, California; James Litton, a senior marketing major from Portland, Oregon; and Alyssa Sepulveda, a junior in the Wine Studies program from Turlock, California. They perform the full range of activities associated with running a tasting room, both behind the bar and behind the scenes. Each works two-to-three days per week, fifteen-to-thirty hours per week, and are paid. Paul Johnson will bring on a couple more people in the fall. An accounting major is being sought to handle the books.

Sandoval's prior wine industry experience included an internship at this year's International Pinot Noir Celebration.

Litton was hired as an entrepreneurial intern in May to assist Matz and Mitchell in opening the Acorn to Oak. He conducted market research, which included visiting other tasting rooms on 3rd Street and preparing pricing outlines. He also collected data on wine clubs. Litton performed more mundane but essential activities as well, such as selecting furniture for the space.

Sepulveda brings the most wine industry experience of the three. Since arriving in Oregon in 2021 just after turning twenty-one, she worked at Argyle and Sokol Blosser, had an internship at Chris James, and now works at Black Dog Vineyards. She aspires to own her own small winery someday.

"The Oak & Vine Society is the origin point for this endeavor, and so to this day, all of the wines we served here are… from a long list of wineries who [sic] are contributors to the wine program at Linfield," Johnson said. In less than three weeks, he increased the list from seventeen to twenty-two participating wineries. The wines available at the bar are from all over Oregon, not just the Willamette Valley, and will continue to be only from Oregon. Johnson views the team as "stewards of the Oregon wine region."

The soft open list included wines from Yamhill Valley Vineyards, Knudsen Vineyards, Johnson's Satyr Fire, Brooks, Celestial Hill, Varnum, Furioso, Richochet, Björnson Vineyard, and Bailey Family Wines.

A wine club is about to be launched. Oak & Vine Society members will be invited to join and receive additional perks. It will also be open to the public. The first allocation will be in Spring 2024. The team will select each allocation from the hundreds of wines that they will be tasting. Each will contain bottles from several wineries.

In addition to the tasting flights, there is wine by the glass or bottle. The latter is also available to take home. Currently, there is one charcuterie board available, but they may expand to something like a humus board. Patrons can also bring in food from the outside to enjoy with a glass of wine.

The comfortable space is well-lit with plenty of windows and is tastefully decorated and furnished. The tasting room has thirty-eight seats inside and thirty seats outside. For overflow or special tastings or events, there is additional space in the back where RJ Studio photography, which shares the location, operates.

Winery takeovers are planned. "I already have several partners who are interested to commit," acknowledged Johnson. These would be either on Friday evening or all weekend.

From its conception, Acorn to Oak was envisioned as a venue where the Linfield community beyond the Wine Studies Program could find a role. The cheerful mural gracing the wall, to the right of the entrance to the tasting room, was painted by art major Allison Hmura. There is also a small gallery with a display of art. "We're going to be constantly showcasing student artwork from Linfield, all of which will be for sale here. We have [Hmura] doing the curation," Johnson noted. The Music Department will be called on to support events that will be hosted in the space. Those with a poetic bent will be invited to express themselves on the menus.

Johnson revealed that "we have also bonded a winery in Linfield's name. So we have an alternating proprietorship…we have a winery operating out of Celestial Hill winery." The first wines to be made this year will be a pinot noir and a chardonnay. Wynne Peterson-Nedry will do the crush and fermentation, with Johnson serving as consulting winemaker. "[It is] not yet known how we will brand the wine—we intend to have rotating art on the labels selected from a student art competition," said Johnson.

By launching the Acorn to Oak Wine Experience in an off-campus commercial area as a learning opportunity for more than just those in the Wine Studies program, Linfield has raised the bar above similar enterprises at other institutions of higher learning.

Follow the events and verify the hours open at Acorn to Oak on its website, acorntooak.com, or on the usual social media. Just drop in; reservations aren't needed.

Heavenly Juice

For Chris and Melissa Thomas, choosing a beverage based on price led to a dramatic change in their careers and lives. Growing up very modestly in Buffalo, New York, Chris never drank, though his mom kept boxed wine. On the rare occasion the family went out to dinner, they only drank tap water and weren't allowed to order anything else.

In 1996, Chris and Melissa were transferred to Belgium. The couple and their infant son would frequent a bistro near where they were staying. Money was tight, so Chris would try to order tap water. He recalled: "And I kept getting a bottle of Evian or bottled water… and the bottled water…was more expensive than the carafe of table wine. And so after like the third or fourth time attempting to get tap water, and I learned what was going on, I told Melissa, 'you're drinking wine.'"

They returned to Belgium for work in 2002, with "a little bit more money so we [could] now afford …to travel across Europe," Chris said. The couple's exposure to wine led to a greater appreciation. They went to France frequently.

"We fell in love with Burgundy …as opposed to Napa and Bordeaux because I grew up a blue-collar Buffalo dude, not fancy, and I love the farms-centric approach to winemaking in Burgundy. It wasn't about the fancy tasting rooms. It wasn't about the chateaus. It was about going to small boutique vineyards and farms and meeting very authentic people where wine was made in the vineyard. And that kind of became …our passion that we really started studying."

Though they considered getting a second home in Burgundy, Chris, a certified public accountant at the time, looked at the real-estate situation, considered exchange rates, and took a different path: "Loving Pinot and Chardonnay…we came to the Willamette" in 2014. The 2012 vintage was being released "and [we fell] in love with Willamette. We said, 'Oh my gosh, like, the heck with Burgundy'… because we've got the great cru, we've got great Pinot and Chardonnay. Absolutely beautiful, beautiful topography, geography, wonderful people."

So, in 2016, the couple decided to focus on Willamette. Waiting until their youngest child finished high school, in 2020 the Thomas's purchased a fifty-two-acre property that included a home built in the 1970s. They renovated it, added a tasting room, and moved from Plano, Texas, that same year.

And what a lovely property they chose. Tucked into the foothills of the Oregon Coast Range, one of the westernmost winegrowing vineyards, Falcon Glen, was planted around 2000, with elevation ranging from 375 to 675 feet. The soils are marine sedimentary and volcanic. After admiring the star-filled dark night sky featuring the Milky Way, the couple renamed the vineyard Celestial Hill. The twelve acres are divided into four vineyards that are farmed organically and named for the Thomas's three children—Benjamin, Kendall Grace, and Brady—and Glenn, in honor of Glenn Howard, the previous proprietor. The vines are under the care of NewGen Vineyard Services.

Chris explains: "These four different vineyards all taste very, very different. And that's what we love… Brady, the highest, at 675 feet, 100 percent volcanic soil, definitely more acidic depth and more about the strawberry and raspberry flavors. And then the lower [vineyards], the Benjamin and Glenn, [are] a little bit warmer. For example,… more of that darker fruit…but none of them are big, heavy wines…

And we've our biggest vineyard, Kendall Grace, which has some Pinot and some Chard. [From the north] section [which] is the coolest on the vineyard… we make our sparkling and our rosé."

Chardonnay was purchased from Fennwood Vineyard in the Yamhill-Carlton American Viticultural Area (AVA) and Anahata Vineyard in the Eola-Amity Hills AVA.

The property is home to chickens, baby-doll sheep, and baby goats. Chris jokes. "We needed thirty-four animals to replace the three kids." I asked Melissa, who worked as a speech pathologist, if she was teaching them to talk. She said no. I guess she's too busy. The Thomases will be adding bees and providing honey to their club members.

In addition to Chris and Melissa, the team includes Jay Somers, the winemaker, and Lita Consoli, the tasting room manager who, like the Thomases, does a bit of everything. Somers, former owner of J. Christopher Wines and current proprietor of J. C. Somers Vintner, recently started as a winemaker at Anne Amie and will be passing the baton to the new winemaking team of Wynne Peterson-Nedry, Alisa Le, and Chris.

The current vintage on offer, a 2020, is the first for the label. There is also a 2021 Rosé of Pinot Noir from Kendall Grace Vineyard that replaces the sold-out 2020. In mid-January 2023, the team gathered at the winery, behind the tasting room in Carlton, to do blending trials of the 2021 vintage wines.

Chris notes: "We do give our wines time. We're not going to rush things…even our '21 Chardonnay, we have not released those… they're resting still in the bottle, and we don't rush anything to market. …We definitely let our Pinots…go eighteen months in the barrel, and then let them rest for a few months in-bottle, and [then] we'll release our '21…We may release one [Pinot] this spring …but we'll leave three to four Pinots for the fall."

I first went to Celestial Hills Vineyard in mid-July 2022 as part of the required visit for inclusion in the *Slow Wine Guide USA*, which was released in mid-2023. At that time, I tasted seven wines from the 2020 vintage, the rosé of pinot noir, three chardonnays, and three pinot noirs. When I returned six months later for this piece, I sampled the same wines except that the 2021 rosé replaced the 2020.

The rosé is intentionally made like a white wine with only the faintest hint of rose-gold. It is different from other white pinot noirs in that the palate is not as viscous but crisp like the darker rosés that are widely available.

The 2020 Eola-Amity Hills Chardonnay from Anahata Vineyard displayed vivid aromas of pineapple and a richer palate. In contrast, the mouthwatering 2020 Yamhill-Carlton Chardonnay from Fennwood Vineyard seems to have become more delicate on the saline palate. The delicious Meursault-like 2020 Barrel Select Chardonnay, also from Fennwood Vineyard, retained its rich nose but is more subtle and nuanced than my earlier notes indicate.

Three 2020 pinot noirs were tasted. The Estate, a blend of vineyards, exhibits floral and red fruit on the nose while the Brady Vineyard is more complex and perfumed. From the dreaded year of the smoke, both are showing some of the impact on the nose and in the texture and should be consumed sooner rather than later. The elegant, dark-fruited, spicy Benjamin Vineyard, on the other hand, appears to have escaped the taint and remains youthful.

In an attempt to avoid smoke taint in 2020, Chris destemmed the grapes, but he did some whole-cluster fermentation in 2021. Because of concerns about inadequate lignification of the stems, no whole-cluster fermentation was done in 2022.

Like many in the valley, Chris regards 2022 as a miracle vintage. Initially fearing a loss of 90 percent of his crop, Chris marveled: "Mother Nature wasn't settling for a bad year. And she came out with a vengeance. And we ended up doing great." They brought in 2.47 tons per acre, which is almost exactly their target. The 2022 vintage is the first one when there will be chardonnay from estate vines that had been grafted over.

Future plans include planting two more acres of Dijon 777 pinot noir and building a winery on the premises. Production goals are thirty-six tons of fruit, with thirty from the estate and six of chardonnay from Fennwood and Royer Vineyards, the latter in Eola-Amity Hills AVA, yielding a total of about 2,200 cases. When asked if there are plans for other varieties, Chris was emphatic: "Never." There is also Méthode Champenoise sparkling wine in the works.

You can enjoy your very own piece of heaven by either visiting the tasting room at 258 N. Kutch Street in Carlton, Wednesday to Sunday, from 11:00 a.m. to 5:00 p.m., or by booking a vineyard tour and tasting. Celestial Hill Vineyard is located at 1970 NW Garris Lane in McMinnville, but don't show up without an appointment.

Monopole Wines distributes Celestial Hill in Texas. Chris is looking into placing his wines in Georgia and some Midwestern markets. The Thomas's go-to restaurant, Tina's in Dundee, carries two of their wines. Two wine clubs offer discounted access to the wines and complimentary tastings and tours. Details can be found at www.celestialhillvineyard.com.

Ground has been broken for the winery on the estate. The tasting room and interim winery will be moving to 3rd Street in McMinnville.

"Rachising" Up Pinot Noir

Stems Add More Dimensions to Willamette Valley's Most Produced Wine

T hink of a grape stem as the skeletal structure of the cluster. The rachis is the spine connecting the cluster to the vine shoot. From it, the peduncle branches off and supports the pedicels, each of which holds a grape. With the advent of mechanized destemmer-crushers, whole-cluster fermentation, which had been the standard method of producing wine for those unwilling to pull the grapes off by hand, became less prevalent.

But an increasing number of pinot producers in the Willamette Valley have gone back to the old way, fermenting with some or all whole clusters for at least part of their offerings. And one even adds oxidized stems from which the grapes have been removed back into the tank. In the same way that cooking fish whole with the bones adds additional dimensions to the flavor over filets, whole-cluster fermentation adds complexity to the finished product that complements and expands beyond what the fruit provides.

What Stems Add

Like the grapes they support, stems go through a ripening cycle during which the chemistry changes. Most winemakers who employ whole-cluster fermentation will only do so when the stems lignify or harden, which diminishes methoxypyrazine compounds that contribute green notes to a wine. Lignified stems add spice, floral, woodsy, and other savory notes to the aroma and flavor as well as silky tannins.

Despite raising the pH and lowering the total titratable acid, or in other words, decreasing the intensity and quantity of acid, tasters report that whole-cluster fermented wines taste fresher. Methyl salicylate, found in stems as well as grape skins, contributes green and minty notes and is likely the primary source. Because stems have a high water content, the alcohol in the wine is slightly lower.

A Few Who Embrace Whole-Cluster Fermentation

Steve Doerner, who retired from Cristom Vineyards in December 2022 and is now winemaker emeritus, is recognized as an early proponent of whole-cluster fermentation in the Willamette Valley. His interest in this approach was kindled when he joined Calera Wine Company in 1978. Josh Jensen, the owner of the California pinot producer, had spent time in Burgundy working with Jacques Seysses, founder of Domaine Dujac and advocate of whole-cluster fermentation. In 1992, Doerner joined Cristom. Following many trials, he decided to aim for 50 percent whole cluster for most vintages.

The first commercial release of Cristom's Whole Cluster Series Pinot Noir is from the 2021 vintage. In December 2022, I sampled the three bottlings, each of which has a different percentage of whole clusters used in the fermentation. The 0 percent whole cluster showed cherry fruit with spice notes on the nose and a tight palate with fine tannins. Lovely floral aromas with some fruit in the background emerged from the 50 percent whole-cluster version. The immature palate had hints of fruit and spice retronasally. The most complex but also the most undeveloped was the 100 percent whole cluster that suggested spice, fruit, and flowers.

Doerner's success influenced others in the valley. Doug Tunnell, owner and winemaker of Brick House, made his wines at Cristom in 1993 and liked what Doerner had achieved. Now he uses "nearly 100 percent in varying proportions from tank to tank." Unlike most other winemakers, Tunnell has used a significant amount of whole clusters in cool years when the stems don't ripen.

"What I found with both the cool, wet, and late 2010 and 2011 vintages was that what might otherwise be excessively light, one-dimensional Pinot Noirs were improved with the structure, tannin, and aromatic profile of 40-to-50-percent whole clusters in the tanks. I can't explain it in any enological detail but simply from my experience with the wines…" he reports.

In contrast, Ken Wright, proprietor and winemaker of Ken Wright Cellars and friend of Doerner, eschews stems in cooler years but will add up to 25 percent in warm years to add freshness.

Brian O'Donnell, proprietor and winemaker of Belle Pente, also cites the work of Doerner and some Burgundians as the reason for experimenting with whole-cluster fermentation. Since 2014, he has used about 15 percent. His lovely 2019 Riona's Block Belle Pente Vineyard Pinot Noir was fermented with 40 percent whole cluster, giving it a floral over fruity nose and a richer palate.

Marcus Goodfellow of Goodfellow Family Cellars was also inspired by the success of Cristom and a few producers in Burgundy and now employs mostly 100 percent whole clusters in his pinot noirs. The youthful 2019 Heritage No. 15 Pinot Noir, fermented with around 80 percent whole cluster, has delicate aromas of stem, flowers, and fruit and a nicely balanced and well-structured palate.

Josh Bergström, director of winemaking at his eponymous winery, concludes, "After twenty-three years of experimentation, we have just decided to ferment all of our Pinot noirs with whole clusters and typically aim for 100-percent whole-cluster inclusion rather than percentages below that." His 2021 Le Pre du Col Vineyard Pinot Noir from Ribbon Ridge was fermented with 100 percent whole clusters as were all of his pinots that year. It displayed pretty floral, sour cherry, and earthy aromas with hints of mushroom. The palate, while young, led with sour cherry and was rich and creamy, with cleansing acidity, obvious tannins, and pleasant texture.

I became aware of whole-cluster fermentation of pinot noir when I started working in the tasting room at White Rose Estate in 2012. Since 2011, owner/producer Greg Sanders has employed 100-percent whole-cluster fermentation for almost all pinot noir.

From 2008 until his untimely death in 2018, Jesús Guillén was winemaker. He credited Gary Andrus with teaching him the fundamentals of whole-cluster fermentation when he consulted at White Rose. Even in the cool, late 2011 vintage, when stems didn't ripen, all wines were fermented with 100 percent whole clusters. The 2011 White Rose Vineyard I sampled in August 2023 put forth an intricate harmonious bouquet with spice, dried dark cherry, green notes, and a hint of forest floor. In contrast, the delicate silky-textured palate revealed elusive flavors of citrus and flowers while lingering long. An elegant example of a cool vintage pinot.

Operations Lead Greg Urmini reports that White Rose is no longer following a rigid protocol but instead "will identify vineyards that will be whole-cluster fermented to achieve our objectives. A way for us to identify a vineyard that could meet whole-cluster criteria could be the elevation…[Objectives include] hold[ing] the body together with more nobility, [making] significant contribution to aromatics, [and] with our neoclassical objective, providing the historical perspective/technique the Burgundians used to create the Pinot Noir breed standard."

Guillén adopted 100-percent whole-cluster fermentation when he founded Guillén Family Wines (GFW). With his widow, Yuliana Cisneros-Guillén, at the helm, the practice continues. "Guillén is 100 percent whole cluster made the way that Jesús taught me," confirms Anthony King, general manager of The Carlton Winemakers Studio, who now makes the wines. "Jesús was really talented at creating elegant, complex, balanced wines with lovely texture, with thoughtful extraction at the end of fermentation," he adds.

This elegance comes through in the nose and palate of the 2014 GFW Damián Winemaker's Cuvée. A predominantly woodsy bouquet is inflected with notes of cherry, roses, and spice. Surprising for such a hot vintage, the palate is refined with great texture and ample structure to support longer aging.

The 2019 GFW Pinot Noir from Domingo Vineyard, planted with cuttings taken from the White Rose Vineyard, has a perfumed nose with fresh herbs, cinnamon, faint floral notes, and a suggestion of citrus. The youthful palate has high acidity, silky tannins, and a smooth texture.

Emerging brightness is the dominant feature of the 2021 GFW Meredith Mitchell Vineyard Pinot Noir with cinnamon, earth, minty chewing gum, and citrus aromas. Though immature, the wine is nicely balanced with a citrus palate, smooth texture, and medium finish.

The studio King oversees is the home to sixteen vintners, all but two of whom use some percentage of whole clusters in at least some of their pinot. "I use whole cluster on about half of the wines that I make these days. Some of my clients prefer pure fruit expressions while others like the complexity and texture of whole cluster. Pinot noirs from my Ratio brand are often 25-to-50 percent whole cluster, with some vineyards fermented without whole cluster and others in the 50-to-100-percent range," explains King.

Both the 2019 and 2021 Ratio Retina Pinot Noirs were still quite young. The 2019 showed more on the nose including dark fruit, cinnamon, and floral notes amidst the otherwise savory aroma. The palates were grippy, with the older wine having a nicer mouthfeel and a hint of caramel. Each has the structure for a long life and promises to blossom into classic examples.

Todd Hansen is grower/wine peddler at Longplay Wine and makes all of the wines from Lia's Vineyard in the Chehalem Mountains. He tells how he came to do whole-cluster fermentation:

"I've been intrigued by whole cluster since I read John Winthrop Haeger's *North American Pinot Noir* back in 2004 or 2005…[which] detailed the fermentation practices and wines of about a hundred wineries, and I loved the descriptions of the whole-cluster wines.

"In 2010, I began selling fruit to White Rose Estate, and I was fascinated by the late Jesús Guillén's beautiful wines. He would just put my fruit in his tank and let it rock… Then, during the 2016 harvest, when we were making my wines with Jay Somers at J. Christopher, Nicolas Potel paid a visit from Burgundy. He said he had a fancy destemmer at his winery but had never used it because he loved whole cluster so much. So, I was thinking I needed to make a whole-cluster wine…

"In 2015, I sold some 828 clone Pinot noir to Willful Wine and Pam Walden. I pitched in on the sorting line and watched how Pam made her wine from my fruit. She went 100-percent whole cluster into the fermenter, which seemed very bold. Then, a year later, in partial payment for the fruit, she traded me two barrels of the resulting wine, and I bottled it and called it 'Experience.' It is a huge wine for a Pinot noir! I love it for its grippy tannins and amazing bouquet. I still have a couple cases of the wine available, and it is still delicious.

"In 2017, we did a fermenter of 100-percent whole-cluster Pinot noir and bottled it up as 'Experience.' We've made 'Experience' every year since (with the exception of 2020, when we made a single bottling of Pinot noir but included our 100-percent whole-cluster fermenter in our 'Explicit Lyrics' Pinot noir)."

In August 2023, I tasted 'Experience' from the 2015 and 2017 to 2022 vintages as well as 2021 'Superstition,' a blend of the rare Mariafeld clone fermented with 100-percent whole cluster and destemmed 115. Each was to varying degrees still young. The 2017 includes two barrels of 100-percent whole-cluster fermented wine, one barrel of 50 percent, and one destemmed. The nose was initially bright, then turned serious, with mint and earth. The beautifully textured palate echoed the mint and also showed fruit, graham cracker, and plenty of acidity. The nicely structured 2018, made from four barrels of 100-percent whole-cluster fermented juice and two destemmed, was intensely minty aromatically and on the palate.

Some other wineries in the valley that add a percentage of whole clusters to a portion or all of their pinot noir fermentations are Abbott

Claim, Archery Summit, Big Table Farm, Brooks Wine, Celestial Hill Vineyard, David Hill Vineyards & Winery, Domaine Divio, Domaine Drouhin Oregon, Dominio IV, EIEIO & Company, Flâneur Wines, Maysara Winery, Raptor Ridge Winery, REX HILL, Ridgecrest, Seufert Winery, Silas Wines, The Eyrie Vineyards, Willamette Valley Vineyards, Winderlea Vineyard and Winery, and Winter's Hill Estate. This list is by no means exhaustive, so the interested reader should inquire as to whether this method is used when out tasting.

A Novel Approach to Stem Inclusion

Rollin Soles, cofounder and winemaker of ROCO Winery, developed a method of getting the goodness from the stems without the risk of greenness that can happen with whole-cluster fermentation when they aren't completely ripe. He explains:

"I came up with the original plan during the 2011 vintage… My motivation was to extract the good things I like about 'whole-cluster' fermented Pinot noir (spice flavor, middle-weight tannin, forest-floor accent), without the bad (greenness, mezcal aromas, broccoli, etc.). Cold whole clusters of Pinot noir are destemmed, then 'restemmed' seven days later! We basically add back as close to 100 percent of the stems as possible. The stems OXIDIZE over a period of seven days [under inert gas] (they are not dried and not lignified)." He calls his approach the "thinking person's use of stems."

As a homage to his time in Australia where stems are called "stalks," Soles labels the resulting wine "The Stalker." The 2017 edition was made from grapes from Wits' End and Marsh Estate Vineyards and aged in neutral oak to avoid any confusion from new barrels. The nose shows spice over red fruit, initially, with herbal freshness, chewing gum, clean mulch, wood, and plummy, dark-chocolate cherry emerging with air. The palate is plush, elegant, with a long spicy finish, excellent balance, and great texture.

The youthful 2021 has very pretty aromas of flowers, spice, and bright fruit. Though immature, the palate is extremely harmonious, well-balanced, tightly integrated, and complex. A textural delight.

Bottom Line

For an increasing number of winemakers in the Willamette Valley, "rachised"-up pinot noir has become a regular part of their offerings. Each has addressed the challenges of dealing with stems that might not be sufficiently lignified to avoid greenness in the finished wine while others prefer to destem all the time. The fact is that beautiful wines can be made with or without stem inclusion. But, if you are on a quest for wines with savory elements and exquisite texture, consider those fermented with whole clusters or spiced up with oxidized stems.

For a more technical discussion of whole-cluster fermentation of pinot noir in the Willamette Valley and additional examples, see my article at https://www.guildsomm.com/public_content/features/articles/b/neal-hulkower/posts/whole-cluster-pinot-noir-willamette-valley.

PART V

Personalities

Remy Wines Takes to the Street for a Fair Celebration of Its Tenth Anniversary

Two days before the street fair celebrating the start of her labels' second decade, Remy Drabkin is on hold with UPS. A shipment of logo shirts commemorating this major milestone went missing two days earlier. With her phone at the ready, the self-described "first Jewish lesbian elected official making wine in the Willamette Valley" (could it be in the world?) recounted the significant events of the last decade. (Remy now identifies as queer.)

During her first vintage in 2006, nine tons of grapes were processed including some pinot gris. Since she had intended to restrict her eponymous brand to Italian varieties, she simultaneously launched Three Wives, which has become her fun label showcasing "really good everyday wines." The year 2011 saw the formation of a wine club, Bell'Amici, (I was a charter member); her family's Lone Madrone vineyard's first delivery, a small amount of grapes for a Three Wives Pinot Noir Nouveau named after her father, Jules; and the release of Three Wives—the Movie to explain the origin of the name. Lone Madrone's first real vintage followed the next year, which also marked the debut of the "baR" (pronounced R bar), a tasting room, gallery, and comfortable hangout.

Every year, fifty to two hundred fifty cases each of sangiovese, barbera, dolcetto, nebbiolo, and lagrein are bottled under the Remy Wine label. Native to South Tyrol, Italy, and grown in Alto Adige for over six hundred years, lagrein is an esoteric variety that Remy has been drawing attention to since her acclaimed vintage 2006 bottling. She now sources all of her lagrein from the estate Lone Madrone vineyard. But her largest bottling is under the Three Wives label, with over eight hundred cases of the 2014 Remy's Red. The one-woman operation has grown to support a six-person staff. Clearly not having enough to do, Remy successfully ran for McMinnville City Council in 2014.

The fair, suggested by Remy's then wife, Laura Pedroni, was held on Sunday, July 17, 2016, on NE 10th Street (how appropriate!) in front of the winery. Some three hundred signed up in advance for admission. During the three-hour celebration, no sense was left unstimulated. Hugs and handshakes were exchanged. Artwork filled the walls and booths while movies featuring the host were shown inside. Pura Vida and Nick's provided tasty sustenance.

Remy Wines and Three Wives, along with Twist Wine Company's Basket Case Wines, a custom crush by Remy, lifted spirits. A standout was the tasting of 2006, 2008, 2010, and the newly released 2013 Remy Lagrein that clearly justified Remy's commitment to the variety. While naturally displaying varying degrees of maturity, each was refreshing on the palate and food-friendly. Portland Opera a la Cart filled our ears with gorgeousness. Flowers and produce lent a farmers market flavor. Games and raffles added a carnival dimension.

Over the next decade, Remy has no ambitious plans, just constant refinements and improvements. With her current space filled and no interest in moving, she isn't expanding production. She simply strives to be a better version of herself.

Unfortunately, the logo wear didn't arrive in time. When it finally does, could this require a second celebration? And if it did, could it beat the first one which was so McMinnville, so much fun, and so Remy?

Remy Drabkin is an indefatigable force of nature. Since this report appeared, she became the mayor of McMinnville in 2022 and served through 2024; added a third label, Black Heart that includes a Willamette Valley chardonnay, a traditional method sparkling wine, and various blends involving Bordeaux varieties; relocated her tasting room to a 1900s farmhouse and built a funky-chic winery that incorporates the Drabkin-Mead Formulation, a carbon-neutral concrete formula, on the Three Wives vineyard (formerly Lone Madrone vineyard) at the foot of the Dundee Hills.

VIDON to Release Space Exploration Series

Former Rocket Scientist Celebrates his Earlier Career

For Don Hagge, space isn't the final frontier. The land is. This one-time space scientist began life on a North Dakota wheat farm and, after a multifaceted career that included stints at NASA, planted an Oregon vineyard. In 1999, he established VIDON Vineyard in the Chehalem Mountains American Viticultural Area with his wife, Vicki, branding it with a contraction of their first names and pronouncing it *vee-dohn*.

The boutique winery above the 12.5 acres of vines produces two thousand to two thousand five hundred cases a year of mostly pinot noir with lesser amounts of chardonnay, pinot blanc, pinot gris, viognier, syrah, and tempranillo, all of them estate-grown as of 2014. Both the vineyard and winery are Low Input Viticulture & Enology (LIVE) certified and participate in the Carbon Reduction Challenge.

In November 2016, Hagge and I were brought together by Carl Giavanti, a winery marketing and public relations consultant, who was struck by our common background as "rocket scientists." After two years flying Navy planes in Korea, Don earned a PhD in physics from the University of California at Berkeley, as well as an Executive MBA from Stanford.

He is much more hands-on than I am. At NASA Goddard Space Flight Center in Greenbelt, Maryland, he designed experiments for Explorer 38, 39, and 40 and then became chief of the Apollo Physics Branch at the Manned Space Flight Center (now Johnson Space Flight Center) in Houston.

On the other hand, my doctorate is in applied mathematics, with a specialty in celestial mechanics—an analytical pursuit that I utilized to design interplanetary trajectories at the Jet Propulsion Laboratory in Pasadena, California. In contrast to Hagge, I am no "metal-bender."

At eighty-five, Hagge is almost a generation ahead of me. As such, he met many of the first American astronauts. In comparing notes, we both knew Karl Henize, one of the original scientist astronauts. As a freshman astronomy major at Northwestern, I was a research assistant to Henize, who had a groundbreaking experiment on Gemini flights 10 to 12.

In August 1967, Henize was named to the astronaut corps. It wasn't until July 1985 that he got into space at age fifty-eight, becoming the oldest person to do so up until that time. Hagge was visibly moved when I told him that Henize died in 1993, attempting but not succeeding to become the oldest person to climb Mount Everest. He still remains in contact with several former astronauts, tasting wine and exchanging e-mails with arcane subjects like "Emergent gravity"

Hagge worked at NASA Houston from the time of Apollo 7 through Apollo 13, after which he had a varied career in high tech that included successful serial entrepreneurial ventures. Before settling in his home in Newberg next to the winery, he lived in various cities in California, Idaho Falls, Seattle, and Portland.

The impending launch of VIDON's Space Exploration Series was the impetus for our more recent meeting. The series comprises three bottlings from the 2015 vintage: Apollo Chardonnay, Explorer Tempranillo, and Saturn Syrah. The chardonnay has a deep yellow color with green overtones and polished aromas. It is mouth-filling, with a longish finish. Drinking nicely now with reasonable acidity, it could age longer. Only twenty-five cases were made.

Not surprisingly, both the tempranillo and syrah were still tight though clearly exhibiting varietal typicity. The former had a juicy fruit aroma while the nose of the latter, cofermented with a bit of viognier, exhibited mostly meat followed by spice. Hagge made one hundred forty cases of the tempranillo and 148 cases of the syrah. A Valentine's Day release is planned.

Having grafted his pinot gris and pinot blanc vines to chardonnay, Hagge expects to make an Apollo Chardonnay reserve every vintage. He is focusing on syrah as an alternative to pinot noir, for which there is no space-series bottling.

However, he continues to produce three single-clone pinot noirs, each named for a grandchild, and a three-clone blend that contains 777, 115, and Pommard. Each is nicely balanced, restrained, and elegant even in hot vintages. There is also a mélange pinot noir that includes the three clones plus Gary Andrus's suitcase clone, AS2.

As further evidence that, as time passes, we all become more of who we are, Hagge remains kinetic, as Vicki is fond of noting. He jumps on his tractor to mow and till as much to relax as to get the work done. He finds outlets for his inventiveness and iconoclasm.

"I'm a scientist, not a winemaker; therefore, I'm not hung up on winemaking traditions," he reminds us. Hagge seals his bottles with glass stoppers and screwcaps, even building his own bottler to accommodate the former. He prefers translucent, food-grade polyethylene oxygen-permeable tanks to stainless steel for aging and argon to nitrogen to displace oxygen. He built his own dispenser for the tasting room that uses argon to preserve wine for weeks. To reduce exposure to oxygen, he designed a bunghole/barrel aspirator with two tubes, one through which argon floods the ullage and provides pressure to push samples through the second. He hopes to try it out soon.

Hagge's skills aren't limited to hardware. He wrote the software for an online meta wine club that allows folks to purchase from several wineries without the usual constraints. He calls it "Vin Alliance" and expects to go live soon.

Although no longer involved in the space business, Hagge and I have remained true to our respective proclivities. While I continue to be paper-bound, writing about wine rather than making it, Hagge applies his talents to making good things for us to drink. VIDON's Space Exploration Series is a testament to the fact that you may take the scientist out of the space program but you can't take the space program out of the scientist.

Alas, VIDON is no more. Four years after I interviewed Hagge, he sold the winery and now is a retired RV traveler. Like the Apollo program to which Hagge contributed, the winery's name has passed into history. It is now called Compris Vineyard. Following their own proclivities, the owners no longer produce the Space Exploration Series. You may not be able to take the space program out of the scientist but you can take it out of a winery.

Jesús Guillén Olvera
(1980 to 2018)

Réquiem para un gentilhombre

Celebrated winemaker Jesús Guillén passed away the night of November 5, 2018, after a short battle with an aggressive form of cancer. He was thirty-eight. His final illness struck as he was cleaning barrels in preparation for the 2018 harvest, one which he would not complete. He was hospitalized in Portland in early October and never returned home.

After earning a degree in computer systems engineering at Universidad Autónoma de Chihuahua, Mexico, Jesús came to Oregon in 2002 to learn English, joining his father, Jesús, Sr., who was working in the White Rose Vineyard. He had no knowledge of or even much interest in wine.

When he began tasting, he was struck by the memories the aromas evoked—a walk in the forest or the floral scents of his grandparents' garden. He named two 1999 vintage pinot noirs his epiphany wines—Archery Summit Arcus Estate and Elizabeth's Reserve from Adelsheim.

Jesús started working with his father in the vineyards and eventually moved to the cellar. Not formally trained, he read every book he could about winemaking, talked to winemakers, and was mentored by Mark Vlossak of St. Innocent, who was consulting at White Rose. Six years after arriving in Oregon, he became the head winemaker for White Rose Estate and was building his own brand, Guillén Family Wines.

During an interview on June 25, 2018, for the Oregon Wine History Archive at Linfield College (https://oregonwinehistoryarchive. org/interviews/jesus-guillen-2/), Jesús explained the stylistic differences between the two labels. For White Rose, "the goal is to portray the classic attributes for pinot noir, so being as pure as we can…the goal is to have elegant classic wines." For his own label, he "use[d] techniques to add complexity to the wine[s], but they have to be in harmony…to evoke memories." At the same time he was becoming proficient in a second language, he was mastering two approaches to winemaking.

Regarding the impact Jesús had on his winemaking, Greg Sanders, owner/producer of White Rose Estate, and its first winemaker, recalled: "He always had a more practical orientation… I would want to do something sort of severe and he would suggest something slightly less severe and we would…. move a little bit more sensibly as a result of it. …I used his tasting skills a lot to…bounce his reactions off of very commonly."

Sanders contrasted their approaches: "Jesús liked to address people's attention to more of the density of the body; he [preferred] a full [wine] and I was more about detail." Sanders acknowledged the support of Jesús and general manager Gavin Joll for adopting whole-cluster fermentation as a significant element of the White Rose style during early trials.

Joll looked back on the twelve years he worked with Jesús: "I will always appreciate his talent, intellect, and ability as a winemaker, but above all, I will admire how caring, kind, and thoughtful he was as a person. He was a truly special person who had a profound and positive impact on those he met." Owing to his familiarity with the procedures and winemaking style, Joll has taken the lead role in creating the 2018 vintage of White Rose wines.

Anthony King, owner of King Wine Consulting, has been providing significant technical support to complete the 2018 vintage of Guillén Family Wines. Jesús's brother, Dagoberto, and his father have also been engaged in creating the wines. In addition, King has been consulting with Joll on the White Rose wines.

Over the past decade, Jesús gained recognition as the first Hispanic head winemaker in the state as well as the producer of consistently high-scoring pinot noir. Earlier this year, he was named one of the "Top 40 under 40 Tastemakers" of 2018 by the *Wine Enthusiast*. He is one of the subjects of the recently released documentary, *Red, White, and Black: The Oregon Wine Story*, about minority winemakers in the Beaver State (http://www.redwhite-black.com/).

As his skills and fame grew, so did his confidence. When asked in the June interview if he believed he makes wines of the same quality as the two wines that changed his career path, Jesús responded, "I think I am performing at that level now." Despite the recognition and his sense of accomplishment, he remained down-to-earth and well-loved. To many, even those who knew him briefly, he was a dear friend. To those in the Hispanic wine community, he became a role model and generous mentor.

Along with Sofía Torres McKay, proprietor and winegrower, Cramoisi Vineyard, and Miguel Lopez, brand ambassador at Domaine Roy & fils, Jesús began forming an organization to recognize and assist Hispanic vineyard workers in achieving their goals in the wine industry. The cover story in the August 2018 issue of OWP (http://www.oregonwinepress.com/breaking-down-walls) featured this project.

Jesús's passing has dealt an anguished blow to *El trio dinamita* as Torres McKay poignantly explained: "There are no words to describe the pain [I feel from] losing our dear friend Jesús, *mi hermano*, so humble, sweet, and the kindest person I ever met, a big fighter with a huge heart, working towards the recognition of our Hispanic workers in the wine industry, working towards a way to make them feel part of our community, [so] that they are not isolated and [know] that we care about them and that we value and appreciate the work they do for us in the wine industry."

Evidencing the resolve to continue, she revealed: "The name of our group will be [*Asociacion Hispana en la Industria Vinicola de Oregon y Comunidad*] AHIVOY as Jesús wanted."

Lopez added: "[W]e will continue to evolve and develop the things that Jesús, Sofía, and myself wanted to see for the future of our Hispanic workforce. We shared the same vision of bridging the worlds that, at times, seemed isolated from one another (vineyard, winery, tasting rooms) and show that we too are as passionate about the work we did, even though we weren't owners, managers, winemakers for the most part. The thought of him not being here to see this to fruition breaks my heart for us and the people we wish to see grow…we are part of the tapestry of the Oregon wine industry, and Jesús was thread and needle….The best way that I can honor him is to keep running forward, don't stop, keep moving."

Torres McKay reported that new members are joining AHIVOY to define and carry out the work.

Jesús is survived by his wife, Yuliana; son, Adrian; mother, Rosario; father, Jesús, Sr.; brother, Dagoberto; and sister, Dalia Duran-Guillén. Members of the Hispanic community, the wine industry, and wine club members joined the family to bid farewell to Jesús at the viewing on November 8 at Macy and Son in McMinnville and the next day at the mass at San Martin de Porres in Dayton and interment at St. James Catholic Cemetery in McMinnville.

A GoFundMe account that was set up to raise money for Jesús's wife and son brought in over $40,000.

Industry response has been reminiscent of the outpouring in 2004 after the sudden death of Jimi Brooks of Brooks Winery, who was also thirty-eight and also left a young son. On November 13, Ian Burrows, owner of Consulon, and Andrew Turner of Valley Wine Merchants in Newberg convened a meeting of some twenty restauranteurs, winemakers, retailers, and media representatives to begin planning a series of fundraising events featuring Guillén Family Wines. Winter's

Hill Estate hosted a wine tasting and sale of Guillén Family Wines on November 24 for the benefit of the family. On December 6, a benefit dinner was held at Ruddick/Wood in Newberg.

During his brief but significant career, Jesús created about a dozen vintages of Guillén Family Wines and ten vintages of White Rose Estate pinot noirs, whose aromas will not only conjure personal memories but, especially for those who knew him, evoke memories of this extraordinary gentleman and winemaker for years to come. Go to https://guillenfamily.com/ if you would like to acquire some beautifully crafted Guillén Family Wines.

This is an article that I wish I never had to write. As I was doing so, especially when watching videos of him shortly after he died, tears were welling up. I still miss him and think about him whenever I open a bottle containing a pinot noir he made.

As you'll read in the next part, AHIVOY is thriving, a living legacy of that most gentle gentleman.

A Post-IPNC Chat with the Master of Ceremonies, Philippe André

Perhaps in anticipation of the 100-plus temperatures during that year's International Pinot Noir Celebration (IPNC), the organizers selected Philippe André, one of the coolest members of the wine industry, as Master of Ceremonies. After a two-year hiatus, due to you-know-what, IPNC was held in person at Linfield University in McMinnville from July 29 to 31, 2022.

André did double duty at the event. In his capacity as US Ambassador, Charles Heidsieck Champagne, he represented one of the over seventy wineries selected to showcase their wines. The day after IPNC and its compact version, Passport to Pinot, passed into history, we sat down for a long chat.

André's career in wine began when he was working at his family restaurant, Oceanique, in Evanston, Illinois. "I saw initially …wine was a language to connect with our guests in a short period of time," he said. "There's something fascinating to me that…we could reach a stranger that has never been to our restaurant and …in five or ten minutes of discussing wine and pouring them their first glass they felt so welcome and at home…"

He then sat in on tastings with distributors after service with his father, the chef. ("By the way I'm nineteen at this moment, so it's a little bit earlier than, you know, maybe I was supposed to be, but we weren't drinking, we were learning.") The "focus to wine really was when I came out to Oregon Pinot Camp in 2013." Later that year, he worked harvest at Maysara. When he returned to Chicago, he joined the auction house, Hart Davis Hart.

Though grateful for "seeing something different from the restaurant world," he was eager to move on. He is now multi-hatted: senior business development manager at Folio Fine Wine Partners, US ambassador for Charles Heidsieck Champagne, and wine director at his family restaurant.

In 2021, André was included in the *Wine Enthusiast's* list of "40 Under 40 Tastemakers," in recognition of his success at Charles Heidsieck in "relaunching and reinvigorating the reputation of the historic brand stateside, as well as developing trade and collector relationships for the portfolio. Through his charisma, passion, and Instagram Live 'Charlie Chats,' there's no doubt that he has fantastically succeeded in this effort." He was even featured by himself on the cover of the issue that contained the list.

This year's IPNC was André's first. "[M]y initial impression was one day I hope I can afford to attend that. [As] a Pinot Camper in 2013, I couldn't wait to come back to the valley and have been back quite a few times," he noted.

He saw his role at the celebration as threefold: "I need to represent the valley here and really show how those wines deserve to be on this international plane. I need to welcome the international folks that travel very, very far to be here to make sure they know that they are very and always welcome here the whole year; and…I need to share and inspire consumers and attendees that joined us [who] either have never been here before or have been many times to show them that we're back…and IPNC is here and it's going to be better than ever."

André also led the Grand Seminar "Through Rosé Colored Glasses: Sparkling Pinot Noir from Near and Far." He was particularly taken with copanelist Pieter Ferreira, the winemaker at Graham Beck in South Africa: "[I]t was a lot of fun to learn more about him and his amazing wines which I have tasted…and now it makes so much sense…the connection between him and the wine, the character that comes in the glass…"

He also admired the contributions of Oregonian copanelists Nate Klostermann, the winemaker at Argyle Winery, and Tony Soter, founder of his eponymous winery. "[I]t was so amazing to try all those different wines and show the audience how unique and diverse sparkling rosé can be and all the different directions the winemaker and terroir can take the wines. I absolutely adore bubbles. I'm a bubblehead. There is no bubble I won't try," he enthused.

Among the memorable wines André sampled during the festivities was a magnum of Joseph Drouhin Montrachet, "some beautiful burgundies, as well," a "hauntingly fresh" 1973 riesling by David Lett, a magnum of 2013 Soter Brut Rosé ("so delicious"), a 2001 and 2002 Patricia Green Singularity, and the 1985 Elk Cove Reserve Pinot Noir. He also had high praise for older vintages from The Eyrie Vineyards. "If these wines can hold up the test of time for ten, fifteen, twenty years and you can show them next to Grand Cru Burgundies, that for me is the seal of approval," he contended.

The most memorable time at IPNC? "I think, honestly, the fun moments that we had at the panel discussions, I won't forget those… [it] was really good to get those guys to feel comfortable to share who they really are, their experience at their winemaking teams, and how they process and produce their wines. I think that is the priority for this event."

But also, "the crass jokes or the banter on the panel? Honestly, it's just so much fun. I wish that we recorded it and made like a highlight reel of them because Tony [Soter] and I [had] some good one-liners going…"

André maintains a distributed eclectic cellar of somewhere around three-to-four thousand bottles. Not surprisingly, it includes a lot of champagne and not just from Charles Heidsieck. "But I'd say half my collection is probably Oregon wines," he revealed. Favorites include Domaine Drouhin Oregon, André Mack's wines, and Brick House. "I'm always buying a ton of wines from Doug [Tunnell]. I mean they're incredible. Doug usually does like an annual magnum sale. So I buy a ton of magnums." He still purchases from his former employer, Hart Davis Hart.

After doing his share to bring down the temperature and to add sparkle to the air at IPNC, this cool dude headed south to chill before returning to Chicago. "I'm gonna go relax in the pool in LA for [a] week."

In the future, André wants to "bring consumers into the realm of beautifully aged champagne at Charles. I've been really focusing on finding younger collectors to…help them decipher this mad world of wine because I know it can be so challenging and intimidating…," he explained. He is also involved in some of the other properties owned by the company including Biondi-Santi, the renowned producer of Brunello di Montalcino, and a sister winery called Rare Champagne, which only bottles small quantities of vintage bubbles. He's been involved in the relaunch of Champagne Charlie, a high-end nonvintage bottling made from as many as twenty-five vintages of reserve wine.

Looking ahead, André is upbeat. "I think the future is extremely bright. You know, we have a dedicated community of folks on the production side that are passionate about making sure and maintaining and raising the bar of quality. [T]he wines that we poured this weekend you can't make without love."

I first met Philippe as he was completing his internship at Maysara in 2013 and getting ready to join Hart Davis Hart. I'll be forever grateful to him for giving me first dibs on some 1949 Volnay-Caillerets from Pierre Latour, from my birthyear, a bottle of which I shared with two dear friends who are no longer with us, John Neatrour (see p. 323) and Jesús Guillén (see p. 230) as part of my 65th birthday celebration.

After failing to connect on several occasions when he came up to the winery I pour at, we finally got together in August 2021 at a small independent wine shop in Evanston called Vinic Wine Co. We were hosted by the proprietor, Sandeep Ghaey, a friend of Philippe, who shared one young and one older vintage of Meursault to wash down the poke bowls we had for lunch. Sadly, the shop shuttered in March 2022 after thirteen years, a victim of the pandemic as well as increased competition in formerly dry Evanston and rising costs. As I wrote in my thank-you to Ghaey, "If your shop existed when I was a grad student, it would have been my second home."

Not surprisingly, Philippe made such a good impression, he was invited back to IPNC in 2024.

PART VI

AHIVOY

Possibly the most consequential piece about the Oregon wine industry I've written, this article has been credited with playing a major role in jump-starting the formation of the Asociación Hispana de la Industria del Vino en Oregon y Comunidad (AHIVOY). I was recruited to write it by Jesús Guillén. It is the cover story of the August 2018 issue of the Oregon Wine Press *and contains stunning photos of the three principals. It was also used to generate awareness and interest among the vineyard stewards.*

Breaking Down Walls

What grows in the vineyards of Oregon? The primary product, of course, is the grapes that are turned into wine. But like those whose labor is essential to the success of the vines and who have been largely unnoticed, there are less obvious yields that are nonetheless of fundamental importance to the industry and the community.

Three members of the Oregon wine industry—two of whom are native Mexicans and one, a first-generation American of Mexican descent—who hold prominent positions are eager to draw attention to the vineyard workers, the majority of whom are Hispanic and ready to advance. Jesús Guillén, Sofía Torres McKay, and Miguel Lopez are in the early stages of forming an organization to increase appreciation of these skilled laborers and to assist them in overcoming challenges to realizing their ambitions. As one part of this effort, each offered their inspirational personal stories and visions for the future.

Jesús Guillén, Winemaker, White Rose Estate, (https://www.whiteroseestate.com/) and Owner and Winemaker, Guillén Family Wines (https://guillenfamily.com/)

Since becoming the first Mexican head winemaker in Oregon in 2008, Guillén has received plenty of press coverage (see: https://www.oregonlive.com/foodday/index.ssf/2013/08/a_pioneering_winemaker.html, https://traveloregon.com/things-to-do/eat-drink/wine-wineries/immigrant-vintners-rise-top/ and https://www.pdxmonthly.com/articles/2018/3/27/jesus-guillen-once-salvaged-cast-off-grapes-now-he-s-got-his-own-winery).

He is now interested in sharing the limelight with the many who are not as celebrated and in serving as an example of what can be achieved when starting out among the vines.

When he came to the United States in 2002 to learn English after earning a degree in computer systems engineering in Chihuahua, Mexico, Guillén knew nothing about wine. Working in the vineyards with his father, Jesús Guillén Sr., who was and still is the vineyard manager for White Rose Estate, and tasting some extraordinary pinot noirs led to an insight that ultimately caused a dramatic shift in his career plans.

"When I came to Oregon is when I saw the connection, not to the wine itself because I did not even know how to appreciate it, but to the vine grape, and we all know where the wine comes from.... [I]f you nurture the vine correctly, it will give you [the] best fruit it can and ...you will have a great wine," he stresses.

Among Guillén's influencers and mentors, Greg Sanders, owner of White Rose Estate, earns considerable praise. "I was very lucky myself to have met Greg Sanders; he gave me the opportunity to become a winemaker and hold the title, which very few Hispanics hold… [He] also taught me how to evaluate a wine with my senses and allowed me to refine my palate by tasting some of the best wines in the world made in Burgundy."

Guillén recognizes "Mark Vlossak, owner and winemaker of St Innocent, who taught me the fundamentals of winemaking, and Gary Andrus, who taught me the fundamentals of whole-cluster fermentation, a technique I have honed to my liking." Reflecting on his strong connection to his family, he credits his grandfather and father for his knowledge of the basics of agriculture and his wife, Yuliana, for her support as he strives to establish Guillén Family Wines.

Guillén's first release under the Dreamcatcher label was in 2007. Not yet owning his own vineyard, he acquires grapes from a number of sources including Vista Hills, Meredith Mitchell, and Domingo, a vineyard on a property owned by Sanders just off Highway 99W and named for Guillén's grandfather.

He is now producing about 1,000 cases a year under the Guillén Family label. Beginning this year (2018), he will be making his wine at the Dundee Hills Wine Library in Newberg. Tastings are also conducted there by appointment, and wines are available for purchase. Guillén Family Wines can also be found at a number of bottle shops and restaurants in the Portland area. In addition, there is a wine club that gets early access to new releases.

Winemaking touches Guillén on several levels. "It…makes you think about the place it was made from, the year it was made, and all the people who made it possible. It is also fascinating because, for me, it satisfies my persistence to achieve something great, something bigger than myself, [it utilizes] my analytical background and, because I am Mexican, wine fulfills my need for challenges. I love to make wine and I love to drink it."

Now his 10-year-old son, Adrián, is showing interest in the wine business and is already helping his father. Guillén shared an utterly charming recording of a tagline he is considering for his brand. Adrián says, "Made in beautiful Oregon" followed by "*por manos mexicanas*" (by Mexican hands) proudly declared in Spanish by his father. Guillén hopes that his son will plan on taking over his label. "[I]t's one of the motivators I have to pursue this business…, to basically create something better for him than it was for me."

Guillén keeps his wine prices affordable while aiming to deliver world-class quality. He hopes that, in doing so, he will be able to attract Hispanic customers who would like to support a Mexican winemaker, as well as others who are simply interested in a great value wine from Oregon.

Despite juggling his full-time job and nurturing his own label, Guillén finds time to mentor the people he works with. He lets them know that "I am trying to pay it forward by making myself accessible. I speak Spanish, and I speak the language of wine. Whoever wants to learn about this, I am here."

The advice that Guillén gives sets realistic expectations. "Know that you will not make a lot of money from this, so the reward you get from working in the wine business is the passion for growing, making, selling, and drinking a beautiful thing. You gotta have passion and love for wine to pursue a career within [this] business. You gotta work hard, learn English if you do not speak it, and learn the language of wine."

Guillén urges us "to change the narrative about who is important in wine. We always speak about the winemaker as the most important person to make the wine but without fruit, there is [no] wine to make… It…does not [just] 'take great people to make great wine;' it takes great people to grow great fruit…"

Sofía Torres McKay, Proprietor and Winegrower, Cramoisi Vineyard (https://www.cramoisivineyard.com/)

When Eugene native Ryan McKay met Sofía Torres in San Francisco in 2001, he introduced her to pinot noir from Oregon. She "fell in love with both." The Mexico City native and consumer of cabernet sauvignon and carmenère became "very curious about the region [and] started drinking more pinot. We started dating and traveling to Oregon wine country, and then we started talking about buying a property and grow[ing] grapes…"

They married in 2005 and bought a ten-acre site on Worden Hill Road in the Dundee Hills in 2011, after selling a house in British Columbia, Canada, in a favorable market. In 2012, they began planting Cramoisi Vineyard at an elevation of five hundred-to-six hundred feet to a mixture of pinot noir clones and now have six acres planted, five in pinot, including the rare clone 122 from Vosne Romanée, a Grand Cru vineyard in Burgundy, and one in chardonnay. The vineyard name means "crimson" in French and derives from the crimson clover covering it in the spring. The McKays are raising two sons in a big house on the edge of the property.

While Ryan continues to keep his day job and helps when he can, Sofía left her position in marketing and sales in 2017 to devote all of her efforts to their venture. Torres McKay is one of the few Hispanic vineyard owners in the Willamette Valley. Since the business is very small, she performs many roles including vineyard management, marketing and sales, event planning, relationship building, management, and administration. She dons boots in the morning and nice clothes at night.

Jessica Cortell, owner of Vitis Terra Vineyard Services, and her crew tend the vineyard, with Torres McKay helping out. Sofía speaks of "the energy of the place" being very important and takes pains to ensure that the vineyard's energy is positive. She can communicate with the workers in Spanish and has developed an easy, informal relationship as well as an empathy with them.

The workers call her Sofía or *la chilanga,* referring to her origins in Mexico City. They joke around. She encourages them to think about their careers and where they would like to end up. "That's something I want to teach and motivate the people in the vineyard, not that you have to be an owner but you can perform on the job." Though they are not her employees, her bond with them is strong. "I'm part of them and they're part of me," she proudly notes.

Torres McKay draws inspiration from a "group of ladies in the wine industry such as Nancy Ponzi, [Susan] Sokol Blosser, Vivian

Weber, [and] Donna Morris, [who] work side by side with their husbands." She also acknowledges the support of several men: "I am inspired by Dick Erath, [who] is my next-door neighbor, Bill Sweat, Jim Maresh, Cliff Anderson, and Moe Momtazi. I just love the way they shared information with me and my husband."

Nancy Ponzi and the others encouraged her to become part of the ¡Salud! Program, which provides health services to winery and vineyard workers. She now leverages her former profession as the only Latina on its marketing committee. She is also active with the Dundee Hills Winegrowers Association. Though, initially, she and her husband faced skepticism about their seriousness in pursuing winegrowing, Torres McKay warmly affirms, "I feel really, really embraced in this community."

Cramoisi Vineyard is farmed biodynamically, though it is not yet certified. In 2014, it produced enough to make one barrel (twenty-five cases) of wine. The following year, 2015, the first formal vintage, also yielded twenty-five cases. Production increased to three hundred cases in 2016 and six hundred cases in 2017. Drew Voit makes the wine with eager disciple Torres McKay looking over his shoulder.

Torres McKay began self-distributing her wine in May 2017 to a limited number of local restaurants and shops. Because of her heritage and gender, prospects frequently ask her about her origins. Some express surprise that she, a Mexican woman, owns a vineyard rather than simply working in one. Initially, she found this exasperating but when she realized that many were only interested in her uncommon story, she decided to make her background a virtue.

She asserts: "I have to make it positive and [use] this [to] my advantage because we are not that many in my position…[I use] myself as a channel to help others and work together as a community."

Cramoisi Vineyard wines are also available through the Crimson Wine Club and at the vineyard, which is open by appointment. A small tasting room will be built in the vineyard. Torres McKay hopes that in ten-to-twenty years, her children will be working in the business and that her wines will find larger markets in the US and Mexico.

The demands of the vineyard, the business, and her family limit her opportunities to "pay-it-forward" by mentoring her workers and forming the group with Guillén and Lopez. To those wishing to enter the wine business, she advises: "[I]t is not easy, but it is freaking great. It is a lifestyle but hard work. Patience is key. Be curious, learn Spanish or learn English, be respectful with the land, and be part of the community and work together to make it better."

Miguel Lopez, Winemaking Production Manager, Domaine Roy & fils (http://www.domaineroy.com/)

For Lopez, interest in making wine the focus of his career developed gradually. "I have always been curious and wondered what happened to the grapes we spent a year growing, then picking and strapping to a flatbed and hauled away. For me, it was never really about a one-bottle epiphany but more of a 'what's next.'"

He was born in 1984—a year not remembered fondly by Oregon winemakers—to immigrants from Oaxaca, Mexico. His father had been the first vineyard manager at Beaux Frères. "He and Michael [Etzel] did all the labor in the early days together and, as a result, as a child, you get dragged along, so I got to see a lot of the Beaux Frères property…and also hang out and work with the family…," he recalls.

While at McMinnville High School, Lopez helped his father, who had moved to another company, doing odd jobs during the summers. Since vineyards were being planted all over the valley, they were active in all of the American Viticultural Areas (AVAs). During the school year, Lopez walked across the street to help at Walnut City WineWorks that was getting started. After he graduated in 2003, he considered joining the Marines, as things were heating up in the Middle East, but was ambivalent.

He approached Zac Spence, owner and winemaker at Walnut City, who responded positively when he asked if he could help with the harvest. This led to his first full-time job in the industry and financial assistance to attend the Chemeketa Eola Northwest Viticulture Center. He obtained an associate's degree in winemaking in 2010 and moved on to positions at Domaine Serene, Soter Vineyards, and Ransom Wine and Spirits over the next two years. He tried his hand at wine sales and marketing during a stint at Vertical Wine and Beer Distribution before joining Domaine Roy & fils in 2013 as they were beginning.

Lopez acknowledges as mentors the two founders of Walnut City, John Davidson and John Gilpin, as well as Zac Spencer, who showed him "that it's hard but also [reminded him] to have fun while doing it." He also credits John Zelko, owner and winemaker of Z'IVO Wines, who helped him through school; Tony Soter, owner and grower of Soter Vineyards; and Tony Gaine and Earl Cramer-Brown of Vertical Wine as major influences.

In 2017, Lopez and his sister, Eva Lopez, formed Red Dirt Vineyard Labor "that takes vineyard workers [with solid skills] and pays them a wage that is above that of a run-of-the-mill farm labor contractor." The starting pay of the workers is where other contractors top off. The wages of the teams are reevaluated regularly and adjusted accordingly.

"It keeps the teams together and shows them that we care about their financial well-being," he says. "We're not a vineyard management company," he insists.

Instead, Red Dirt operates with a different business model. The highly skilled team takes direction from a vineyard manager or consultant and follows his or her practices. The business keeps a very low profile, with no web presence, relying on word of mouth to attract both engagements and team members.

Starting Red Dirt resensitized Lopez to the challenges facing vineyard workers and motivated him to spearhead the formation of the group with Guillén and Torres McKay. With so many issues to address and paths to consider, plans remain inchoate.

"There are a lot of things up in the air right now. The way I see it going is baby steps," he admits. He credits Torres McKay with "rounding them up when they slack off." The three hope that telling their stories is the beginning of a continuing process of drawing media attention, not just to the most visible Hispanics in the industry but to those who labor behind the scenes.

Lopez cited the National Farm Workers Association, which later became the United Farm Workers Union, cofounded by César Chávez, as the only working model for the yet unnamed organization. He acknowledged that the environment these days is much easier than what Chávez dealt with. "I think we're in a really unique situation where we can do a lot in a really small amount of time. [In five years] I see ourselves having developed a really core group of people...that believe in what we are trying to accomplish... [and] want to be involved... I can see it becoming woven into the fabric of what our industry is."

Beyond the raw material that becomes wine, Oregon vineyards are fertile sites for strengthening family ties, gaining skills, building self-confidence, nurturing careers, and advancing the dreams of those who work in them. Guillén, Torres McKay, and Lopez envisage their new organization stimulating these other essential yields that will bring long-overdue changes to the perception of vineyard workers and create opportunities to realize their aspirations.

Torres McKay suggests: "The next time you open a bottle of wine, recognize not only the owner or fabulous winemakers but also the amazing work our vineyard stewards do in the valley for us to grow phenomenal wines. Wave at them and say, *Gracias!*"

There They Go!

AHIVOY Sets Sail

Jesús Guillén must be smiling. The organization he envisioned to help vineyard workers broaden their opportunities in the wine industry is real and has begun affecting lives. *Asociacion Hispana en la Industria Vinicola de Oregon y Comunidad* was conceived and named by the late winemaker for White Rose Estate and Guillén Family Wines with Sofía Torres McKay, proprietor and winegrower, Cramoisi Vineyard; and Miguel Lopez, co-owner of Red Dirt Vineyard Labor. AHIVOY, the acronym by which the group is known, translates from the Spanish to "there I go." And it has gone far since the idea came to light in August 2018.

The Debut

AHIVOY held its first public event on November 3, 2019, at Pura Vida in McMinnville. A fundraising dinner and silent auction attracted forty-two attendees, including winery and vineyard owners, and raised $5,000. Chef Ricardo Antunez prepared a brilliant menu of French-Latin fusion dishes paired with wines from Alumbra Cellars, Grochau Cellars, Guillén Family Wines, and Cramoisi Vineyard.

Organizers announced that AHIVOY was in the process of obtaining 501(c)(3) tax-exempt status and had begun accepting donations to fund the first Wine Industry Professional Training for current vineyard workers. One of the first students, Marco Corado, now at Hyland, was a special guest.

In Formation

Key positions in the association have been filled, with others still open at this writing. The president is DeAnna Ornelas, of Winderlea Vineyard and Winery, and Lopez and Sam Parra, of Parra Wine Co., are chair and co-chair. Parra is also responsible for fundraising and sponsorship procurement. The finance committee includes Karly Tell, CPA, of Irvine & Company, and Yuliana Guillén, Jesús's widow and proprietor of Guillén Family Wines.

Torres McKay; Jessica Sandrock, former director of wine studies and agricultural sciences at Chemeketa and AHIVOY board member; Rich Schmidt of Linfield College; and Elena Rodriguez of Alumbra comprise the education committee.

Marketing and events committee members are Heidi Moore of Country Financial, Stephanie Hoffman of VIDON Vineyard, Sami Márquez Sattva of the Lambda Lion Group, and Ornelas. Their primary responsibility is to provide the resources and guidance to undertake AHIVOY's education-focused mission.

Accomplishing the Mission

The educational initiative is the realization of AHIVOY's mission statement, which was "formed with the purpose of creating opportunities and empowering Latino vineyard workers of the Willamette Valley through education to overcome socioeconomic challenges and strive toward realizing their ambitions. AHIVOY seeks to break down the barriers that many vineyard workers face in relation to education and income disparities."

The organizational goals emphasize creating a career path, encouraging learning English, attracting "more students to continue their education in the wine industry and stay in [it]," and "[collaborating] with the community to promote education at all levels." Torres McKay urges workers to: "Learn something new every year and inspire others to do the same, this will allow you to open the door to new opportunities, be ambitious, be curious, and make your dream job come true."

The first step toward that goal was taken on January 15, 2020, when the Wine Industry Professional Training Program, developed in coordination with the wine studies program at Chemeketa Community College, began as a pilot at its Salem campus. Sandrock was instrumental in designing the English-language curriculum and has since handed the reins to interim Director, Paul Davis, and Program Coordinator, Megan Jensen. Elizabeth Cryan, the training facilitator, will monitor how well the students learn the language and contribute to the decision on whether changes need to be made going forward. The first of two ten-week terms runs through March 18. The students will spend one six-hour day per week in the classroom and on field trips to vineyards and wineries.

During the first term, the students will learn vineyard and winery terms in English, demonstrate basic job-related communications, and obtain a detailed overview of the process of growing grapes and making wine, complying with government regulations and best practices for sanitation, health, and safety.

Instructors for this term are Allen Holstein, a retired vineyard manager; Don Crank, of Hawks View Cellars; Will Hamilton, of Violin Wines; and others from the industry. AHIVOY board members will also share their wisdom.

Running from April 1 to June 3, the second term will focus on wine tasting and sales, with additional exposure to relevant terms in English. Other topics include learning wine labels, providing customer service, wine sales and marketing, tourism and wine, tasting room models, and educational opportunities beyond the program. Instructors are to be determined. Each term has formal measures of success tied to performance-based learning outcomes.

"Other assessments will be facilitated throughout the training to determine if students gain a greater understanding and awareness of the complete winegrowing, making, evaluating, and sales/business process. A major goal of this program is to help students see how their work fits into the bigger picture. Also, to explore additional learning opportunities for career growth and leadership in the industry," explained Sandrock. The successful student will receive a noncredit certificate at the end of the training.

The program's progress will be preserved in the Oregon Wine History Archive at Linfield College. "We are…working with [Director] Rich Schmidt to start interviews with students and vineyard stewards and the wineries that are supporting these students," confirmed Torres McKay.

Assessing and Acquiring the Resources

The cost of the program for the pilot year is $1,200 per student for the two terms. In addition, AHIVOY offsets the wages lost for the hours in class. Students are expected to contribute $300 over two or three payments to encourage commitment.

The Erath Family Foundation gave AHIVOY a grant to fund all the students for 2020. Dick Erath wrote in an email: "AHIVOY is a grassroots group dedicated to the advancement of our Hispanic workforce that is so instrumental to the success of the Oregon wine industry…they deserve our full support."

The First Day

On January 15, the first cohort of students began their studies. They represented Hyland, Lange, Montinore, Ponzi, Results Partners, Vitis Terra, and Wine by Joe. After the welcome by Cryan, Torres McKay, and Lopez, and introductions by the students, Holstein delivered the first lessons on vineyard management.

Topics covered were useful vineyard calculations, vine balance, pruning weight, record keeping, accounting, and commercial aspects of the Oregon wine business. Despite the compressed schedule of the program, the subjects were covered in-depth.

For instance, the students were given handouts that covered conversions between units and rules of thumb for determining key vineyard parameters, such as tons per acre and yield. The class also went out to the adjacent vineyard to learn how to prune.

When asked what their goals for the course were, many who worked in the vineyard wanted to learn more about the winemaking process while others were interested in being promoted to management roles. Sergio Reyes, vineyard manager at Montinore, said, "I want to be an example for my kids."

Award-winning author Katherine Cole hosted an invitation-only fundraiser at her home on February 20. She explained her interest: "When I interviewed Sofía Torres McKay and Miguel Lopez for an article about the labor crisis for *SevenFifty Daily*, I was struck by the elegant simplicity of the AHIVOY concept. It's very straightforward and easily actionable. The more money the program earns, the more workers can enroll in the program."

Looking Ahead

With funds for this year secured, mobilizing support for 2021 is the top priority. Torres McKay "encourage[s] the wine industry to [learn] more about [the program], sponsor a student, send [an employee]… and of course, [contribute] so we can continue our efforts."

Jesús Guillén reminded us: "The skilled manual labor that Hispanics provide in the vineyard is essential for making high-quality wine" and urged us to "Give them opportunities to grow and also to learn more about the beautiful product they are helping create." AHIVOY and Chemeketa are now fulfilling this dream.

For more information, including how to contribute, visit AHIVOY on Facebook, #ahivoyoregon on Instagram, www.ahivoyoregon.org, or email info@ahivoyoregon.org.

The pandemic caused an immediate halt to the program.

AHIVOY Presses On

Less than two months after it began, the Wine Industry Professional Training Program, the brainchild of the *Asociacion Hispana en la Industria Vinicola de Oregon y Comunidad* (AHIVOY), abruptly paused on March 13, 2020, with one week left in the first term, when the coronavirus forced Chemeketa Community College in Salem to close.

In the interim, planning and fundraising have continued in the manner that evidences the determination and resilience that has characterized the organization since its inception less than two years ago. In addition, AHIVOY has been increasing the awareness in the Oregon wine industry of the benefits of the training for their vineyard stewards, the apposite designation adopted for the participants.

Jessica Sandrock, a member of the education committee and coordinator of programs and grants, wrote in an email: "All students are expected to return. Eleven vineyard stewards started the program in 2020. Ten will be able to complete the program. We are sad to share that one of the students, Ben Hernandez (Lange Estate), lost his battle with cancer and passed away in August." When I spoke with him on the first day of class in January, Hernandez had said he wanted to become a manager.

During a Zoom session I convened in October to get an update, Sandrock mentioned that "according to employers [of the participants]…the general consensus was that all of the vineyard stewards were eager to get back and finish the program…" The employers asked the organizers to "consolidate the program a little bit to meet the seasonal needs of when work ramps up."

Montinore reported that its two participants have already stepped up to more responsibility. AHIVOY cofounder Sofía Torres McKay, proprietor and winegrower, Cramoisi Vineyard, who spoke with some of the students, said, "All of them have…very positive feedback. They want to go back. They are also thinking about what they can do after…" If all goes as planned, the first cohort will return in person on January 13, 2021, picking up from where the program had left off, and end on March 3, 2021.

During the last weekend of September 2020, the organization held its first online auction. DeAnna Ornelas, of Winderlea Vineyard and Winery and president of AHIVOY, reported that 117 folks from around the West registered to bid on thirty-six donated lots that included wine packages, private virtual tastings, two-night getaway vineyard stays, and tours of two American Viticultural Areas. The auction raised $18,000.

"We wanted to give the bidders an opportunity to learn, to give them wine-learning experiences, not just give them wine. Our goal is to bring people to the valley and [have them] get to know the people who are making wines, the vineyards, who's in the frontlines," stressed Torres McKay. AHIVOY wants to "do something good, not just for our vineyard stewards but also for our donors," she emphasized.

With the revenue from the auction and other donations, there are enough funds to cover scholarships for the second cohort, said Sandrock. Because the staff are all volunteers, less than 5 percent of the budget goes to administrative, operations, business expenses, and specialized contractor fees, a remarkably small percentage.

The second cohort is being recruited and will include at least one woman. In her email, Sandrock stated: "There are nine interested students right now. However, AHIVOY is still accepting applications, and we encourage vineyard stewards and employers to turn applications in at their earliest convenience. There will be an orientation after harvest for new students and employers. The 2021 cohort will start their program [on] January 12, 2021, and complete the program on April 27, 2021. AHIVOY can award eleven scholarships for the 2021 cohort (the limited number of scholarships this year is so we can meet social-distancing requirements in place at campuses where most classes will take place)."

Rich Schmidt, Linfield University director of archives and resource sharing and a member of AHIVOY educational committee, told of a parallel effort that predates the formation of the association to do bilingual interviews of twenty-five-to-thirty vineyard workers and perhaps assemble them into a short documentary.

Sandrock added that through Greg Jones, director of the Wine Studies Program at Linfield, and Bree Stock, WSET Level I instruction will be included in the second term curriculum of the Wine Industry Professional Training Program, with Stock donating her time to teach.

Also, Chemeketa is "attaching continuing education units to the program," which will automatically convert to credits for the introductory courses at the college and are transferrable to the Wine Studies Program at Linfield, OSU, and Umpqua.

AHIVOY serves as an excellent example for any unrepresented group seeking to increase participation in an industry. It brilliantly balances individual initiative with community support to expose participants to the widest range of options and begin their preparation to assume new roles. It is an organization worth cloning.

As Schmidt noted: "[AHIVOY] speaks to Oregon again being a trailblazer… in the industry. No doubt…other [wine regions] will look at this as a model."

Graduation Day

On a cloudless March 3, 2021, the dream of three commenced dreams coming true for eight. The octet of vineyard stewards (see below) comprised the first class to complete the Wine Industry Professional Training Program established by the *Asociación Hispana de la Industria del Vino en Oregon y Comunidad* (AHIVOY).

In 2018, the late Jesús Guillén, Sofía Torres McKay of Cramoisi Vineyard, and Miguel López of Red Dirt Vineyard Management and Winemaking conceived of an organization that would broaden the opportunities for those who grow the grapes that have established Oregon as a premier wine-producing region.

Joined by members of the AHIVOY board, seven members of the first cohort gathered outside the adorable Cramoisi tasting room, in the Dundee Hills amidst the vines, to celebrate finishing their coursework at Chemeketa Community College in Salem that had begun on January 15, 2020. It was interrupted on March 13, 2020, then restarted on January 13, 2021, in person, in accordance with the school's COVID policies.

The reaction of the members of the cohort was uniformly positive and reflected a strong desire to pursue additional education. Alejandro Moreno of Results Partners acknowledged reinforcing "the motivation to keep learning" as the most important benefit. He hopes to "get a degree in vineyard management."

Sergio Reyes of Montinore Estate and Daniel Barajas of Results Partners want to take General Educational Development (GED) classes and learn more English.

Alejandro Avalos of Montinore Estate praised the "support from the Oregon wine industry [and] instructors" and plans to earn a degree in vineyard management.

Omar Perea of Beaux Frères mentioned his "next step [is] taking GED," and then other classes. Miguel Azua of Wine by Joe "wants to be a winemaker…[and will] take some computer classes, English classes, [and] winemaking." José Martínez López of Ponzi Vineyards found that "everything was pretty interesting."

Several are looking to continue at Chemeketa, which has GED and English as a Second Language classes as well as viticulture and winemaking. Matching funds are available for some of these courses.

Jessica Sandlock, a member of the AHIVOY education committee and coordinator of programs and grants, who has been deeply involved in planning, is "gathering feedback from students, instructors, employers…then working through [it] over the next months." She admitted: "It wasn't easy all the time to complete the program."

Compounding the challenge was the start of the second cohort (see below for members including four women, the first to participate in the program) on January 12, the day before the first cohort came back to class.

"Chemeketa did a fantastic job [handling the two cohorts at the same time] … they really went above and beyond," noted Sandlock. The second graduation ceremony is scheduled for April 27, 2021. Sandlock said that there is funding for twelve scholarships for the 2022 class, with more fundraising events being planned to cover 2023 and beyond. New partnerships are being formed as well.

Before the ceremony, a lunch catered by Red Hills Market was served, accompanied by wines from Cramoisi Vineyard, Guillén Family Wines, Gonzales Wine Company, Alumbra Cellars, and

Parra Wine Company. In addition, Argyle, which donated $5,000 for scholarships, contributed three 2018 vintage wines from its *Ojo Brilloso* (*shiny eye* or *nice observation*, colloquially) program (https://argylewinery.com/about/winemaking/ojobrilloso/). Each was made from grapes harvested from a small pinot noir block in a vineyard chosen, cultivated, and vinified by each vineyard's manager. The vineyards are Spirit Hill, Knudsen, and Lone Star.

From the website: "Contributions from the Ojo Brilloso program are being made to ¡Salud!, AHIVOY, and The Roots Fund. In addition, Argyle will be working with Executive Directors of the AHIVOY and The Roots Fund organizations to identify opportunities for qualified participants to have access to Argyle programs and people with the underlying objective of creating more opportunity for people of color in the wine industry."

With attendees well-fed and moistened, the graduation began. AHIVOY President DeAnna Ornelas of Winderlea Vineyard & Winery recognized each of the vineyard stewards with a few personalized comments, and then presented certificates of completion and goodie bags. Other board members offered additional praise and encouragement.

Sandrock acknowledged the time and dedication required to finish despite the demands of full-time jobs and the pandemic. She appreciated the feedback already given and solicited more on forms.

Elena Rodriguez of Alumbra Cellars commented that each of the participants was pushed outside his comfort zone, and that is how one grows. "Just remember, the energy you put out will one day come back to you," she advised.

López encouraged the group to pay it forward. "Make sure you find that one kid or that one person that's interested and nurture them and help them grow," he urged.

Rich Schmidt of Linfield University was "excited to see what comes next for all of you" and also asked that they help find the next groups for the program.

Torres McKay observed that "this is a dream come true not just for us, not just for you" but for setting a path for future generations.

She reminded them: "You guys know where the wines come from. You guys know where the wines are born and the end…product is thanks to you."

Sam Parra of his eponymous wine company was given the last word. After recounting the highlights of the short history of AHIVOY, he counseled the vineyard stewards: "You should make the rounds in the valley, educate your palate."

The ceremony concluded with each of those who completed the program saying a few words.

Despite the demands of their critical work in the vineyards compounded by the nightmare of the coronavirus, eight men with dirt under their nails have taken a huge leap toward finding their place in the Oregon wine industry as pioneers in their own right, with the support of an association that was only a vague notion less than three years ago. That is a new definition of remarkable.

First Cohort: Alejandro Avalos Corona, Montinore Estate; Miguel Azua, Wine by Joe; Daniel Barajas, Results Partners; José Martínez López, Ponzi Vineyards; Adrian Mendoza, Montinore Estate; Alejandro Moreno, Results Partners; Sergio Reyes Silva, Montinore Estate; Omar Perea, Beaux Frères.

Second Cohort: Cynthia Hernandez, Alumbra; Leydi Gonzalaz, Vitis Terra; Javier Castaneda, Nysa Vineyard; Enrique Cervantez Diaz, Lange Estate; Ivan George, Vitis Terra; Jubencio Nezrete, Vitis Terra; Ulises Ayala, Arlyn Vineyard; Sonia Nieto Solano, Archery Summit; Eva Lopez, Red Dirt; Sam Lopez, Red Dirt; Roman Carbajal Franco, Atlas.

This is a summary of the early history of AHIVOY that appeared in the July–August 2021 issue of The Grapevine.

The Vineyard Stewards' Stewards

Two were born in Mexico and one in the US to parents from Oaxaca. Each had carved a path to success in Oregon's wine industry and wanted to pay it forward by easing the way for those at the beginning of the winemaking process, the vineyard steward. An association they created not that long ago has been fulfilling their vision.

The Trio of Founders

After obtaining a degree in computer systems engineering in Chihuahua, Mexico, in 2002, Jesús Guillén came to the United States to learn English. His father, Jesús senior—who then, as now, was managing the vineyards at White Rose Estate in Dayton, Oregon—put him to work.

Blossoming under the guidance of mentors—including White Rose's owner Greg Sanders; consulting winemaker Mark Vlossak, who also owns St Innocent; and the late Gary Andrus—the younger Guillén quickly moved from the fields to the cellar. In 2008, he became the first Mexican head winemaker in Oregon. He also started his own brand, Guillén Family Wines.

Mexico City native Sofía Torres McKay was working in the technology field when she met her husband, Ryan, in San Francisco in 2001. After they married in 2005, they acquired ten acres in the Dundee Hills American Viticultural Area and planted Cramoisi Vineyard. They bottle estate pinot noir, red and rosé, and chardonnay under the Cramoisi label.

Native Oregonian Miguel Lopez was born to immigrants from Oaxaca and raised in wine country. His resume includes positions at several wineries and a distributor. He now dedicates his time to Red Dirt Vineyard Management and Winemaking, a venture he started with his sister, Eva Lopez, in 2018.

From Idea to Reality

That same year, Guillén, Torres McKay, and Lopez went public with their plans to form an organization named the *Asociación Hispana de la Industria del Vino en Oregon y Comunidad* or AHIVOY (ahivoyoregon.org), which is Spanish for "there I go."

Tragically, Guillén died at age thirty-eight on November 5, 2018, after a short battle with an aggressive form of cancer. His widow, Yuliana Cisneros-Guillén, took his place with the other founders and also maintains the family's label. She promotes the importance of those the group is dedicated to supporting: "AHIVOY vineyard stewards are tending the vineyards that capture our Oregon wine region in every wine that is being produced."

The association adopted an ambitious and sharply focused mission statement: "AHIVOY strengthens the Oregon wine community by empowering Vineyard Stewards through education." It collaborated with Chemeketa Community College's Wine Studies program, at the Eola campus in Salem, to develop the Wine Industry Professional Training Program tailored to the constraints of full-time vineyard workers.

AHIVOY held its first public event in November 2019 to raise funds for this project and to announce that it had begun selecting members of the first cohort. The Oregon wine industry and supporters quickly rallied to the nascent organization. A major boost came from the Erath Family Foundation, which covered the expenses for all students in the inaugural class.

The First Cohorts

On January 15, 2020, a small group of vineyard stewards gathered in a Chemeketa classroom for the first time to expand their view of the wine industry. Over the two ten-week terms, topics covered the entire process from vineyard to glass, incorporating the details of grape growing and vinification as well as tasting and marketing the final product. Along with the rest of the world, the program came to a sudden halt on March 13, one week shy of the end of the first term.

During the forced hiatus, the now tax-exempt association, with officers, a volunteer board, and committee structure in place, continued to raise funds for a second cohort and to recruit students. They successfully accumulated enough to fund the second cohort, which started on January 13, 2021, one day after the first cohort returned to class.

On a cloudless March 3, 2021, the dream of the three founders commenced dreams coming true for eight men, the first to complete the program. The second cohort, comprised of four women and six men, celebrated its graduation on April 27.

Reactions, Initial Impact, and Follow-On

Jessica Sandrock, a member of the AHIVOY education committee and coordinator of programs and grants, was instrumental in designing the English-language curriculum for the program. She collected and shared feedback from students in both cohorts and their employers.

"Overall, the reaction to the program has been really positive," said Sandrock. She added that, not surprisingly, the students overwhelmingly liked gaining more advanced technical knowledge on vineyard management.

One wrote: "Vineyard management classes are very good. [It was c]hallenging, and I learned new things that I am using at work already."

But as they got into winemaking topics, they got interested in those. One student in the first cohort is pursuing winemaking and his own label. Another valued "learning more about all of the things that go into growing grapes and making wine. I will use all of this in my work."

Most enjoyed visiting different vineyards and wineries, learning different ways to train the vines, and the work of the winery. Three members of the second cohort really appreciated the WSET Level 1 training and certification that was added this year, and they will pursue the higher levels. Several plan to continue their education by taking classes to deepen their knowledge of vineyard management, study enology, learn English, or get a General Educational Development (GED) certificate. The respondents unanimously plan to recommend the program to other vineyard stewards.

Employer reaction was also strongly positive. All agreed that "the vineyard steward [is] showing increased eagerness to learn" while 80 percent affirmed that "the vineyard steward [is] demonstrating more versatility."

Jesse Lange, of Lange Estate Winery & Vineyards, wrote: "We've been fortunate to have two very valued and experienced employees participate in the AHIVOY program for each of the two years the program has been available.

"Continuing education has a holistic [effect] on any student—one that has the potential to positively permeate many aspects of job performance. We've seen that [to] be the case here at Lange Estate from the viticulture, wine production, and even sales and marketing. Both Benjamin and Enrique [Cervantez Diaz] have shown higher levels of enthusiasm, deeper levels of questioning, and general happiness with the opportunity to expand a knowledge base and skill set. Also, the chance to learn amongst peers allows for interactions that can [be] cohesive and collaborative—especially coming out of the pandemic. All of this is healthy indeed!

"We would definitely recommend this program to other folks in the industry—no doubt!"

Sam Stetser of Atlas Vineyard Management sent one of his employees to the second cohort. "The great thing about AHIVOY is both myself and Roman [Carbajal Franco] were on board with doing it; it wouldn't work if that wasn't the case." He hopes to "transition Roman into a management role with more responsibility."

Sandrock said that they will track graduates and are trying formal and informal ways to keep them connected to the association as ambassadors or board members. She also stated that AHIVOY is working with Oregon State University to support graduates interested in pursuing a bachelor's degree.

This can be an attractive option since the graduates accrued continuing education credits that can be used to place out of the three introductory courses at Chemeketa in its Wine Studies Program whose credits, in turn, readily transfer to OSU.

More immediate opportunities are with the OSU extension. Discussions are underway with Prof. Patricia Skinkis, viticulture extension specialist, about specific topics she can support, such as pest management.

The Future

Thus far, AHIVOY's reach has only extended to the Willamette Valley, and mostly the north, at that, but there has been outreach to Southern Oregon and The Rocks District. However, more involvement is needed to spread the word and do all of the other critical functions of the growing organization.

Resources for the 2022 class have been secured and applications for membership in the third cohort are being accepted through November 15, 2021. Classes are scheduled from January 5 to April 27, 2022, on Wednesdays, from 9:00 a.m. to 3:00 p.m. In the meantime, fundraising continues to ensure classes can continue beyond next year and perhaps even expand to include larger numbers of students.

In less than three years, the vision of the founding trio has taken hold, gathering widening support from an industry known for collaboration and concern for all of its members. With momentum building, AHIVOY looks to be as successful as ¡Salud!, the organization that has been providing medical services to Oregon vineyard workers and their families for over twenty-five years.

While ¡Salud! maintains the health of the vineyard stewards, AHIVOY enriches their intellect and feeds their curiosity. As founder Torres McKay asserts: "The more we empower vineyard stewards through education, [the more] we will become the best winegrowing community, making the best wines in the world."

PART VII

Slow Wine

Playing the Field for Slow Wine

cknowledging "Oregon's commitment to sustainable winemaking and respect for the terroir… consistent with Slow Wine's principles and its mission to support local agriculture," fifty of the state's wineries were reviewed for the first time in its 2019 Guide, joining California as the only region outside Italy and Slovenia included.

That number grew to sixty-nine in the 2020 guide, with eighty-nine under consideration for next year's release, along with Washington and New York wineries. My second year as one of the field coordinators was 2020.

The *Slow Wine Guide*, an offshoot of the Slow Food movement, first appeared in 2010. Until now, it required its field coordinators to make onsite visits to all the wineries to meet and taste with a principal, preferably the owner or winemaker. In these strained times, the rules had to change.

National Editor of the US edition of the guide, Deborah Parker Wong, wrote: "We will not be making any in-person winery visits this year." Instead, "We [will] record virtual tastings with the winemakers/vintners on Zoom or Skype (or your preferred platform), then share the recordings."

I was granted dispensation since seven of the ten wineries I covered agreed to meet in person with precautions taken. The remaining three tastings were over Zoom.

Before I share my experiences, a word about the winery selection criteria is merited. While formal certification is not required, adherence to sustainable practices is. "We do not include wineries that use synthetic inputs for weed control—i.e., glyphosate—that are not permissible in organic or biodynamic programs, period," stresses Parker Wong. Of course, quality is important, but the quality-value ratio is particularly prized.

Returning coordinators were asked to revisit wineries that they had previously covered. I had done nine and added a tenth. In addition to short write-ups about the people and vineyards, I prepared notes on three of the wines tasted at each. What follows are snapshots of my encounters, with wineries presented in alphabetical order.

I met with Brian O'Donnell at the Belle Pente Vineyard & Winery to taste three of his pinot noirs, the fruity 2017 Murto Vineyard, the aromatic 2017 Belle Pente Vineyard, and the texturally interesting 2016 Estate Reserve. Brian's wines consistently offer excellent value and high quality.

Josh Bergström, of Bergström Wines, and I tasted together on Zoom. I had picked up seven bottles—three single vineyard chardonnays, the first releases of these wines, and four single vineyard pinot noirs, all from the 2018 vintage. Thanks to my Coravin, I was able to draw off samples of each and save the rest. The Bergström Vineyard Chardonnay offered nutty, lemony aromas and a wonderful palate with distinct salinity. The expansive Salice and flowery Le Pre du Col Vineyard Pinot Noirs are also in my submission.

I visited with Pattie Björnson and her winemaker, Scott Sabbadini, at Björnson Vineyard, another source of high-quality-to-price bottles. Of note are the bright 2019 Viola Auxerrois, the creamy 2017 Reserve Pinot Noir, and the floral 2017 Isabel Pinot Noir.

For our Zoom session, Doug Tunnell of Brick House Wines provided the juicy 2018 "Cascadia" Chardonnay, a complex 2018 "Les Dijonnais" Pinot Noir, and the accessible 2018 "Halliday Hill" Pinot Noir, all sourced from the estate biodynamically certified vineyard.

Assistant winemaker Tracy Kendall, hosted me at the interim tasting room of Nicholas Jay. She poured the first release of the hunger-inducing 2018 Affinitiés Chardonnay and three 2017 pinot noirs, including the elegant Willamette Valley and the polished Bishop Creek.

During my stop at the winery, Patrick Reuter, owner and wine-maker of Dominio IV, shared six bottles: a viognier, a syrah, two tempranillos, a chardonnay, and a pinot noir——displaying his broad interest and versatility. The intriguing 2017 Imagination Series Chardonnay from Wahle Vineyard, the pretty 2015 "Sketches of Spain" Tempranillo, and the food-friendly 2014 "Rain on Leaves" Pinot Noir made my report.

This was the first year I covered Kelley Fox Wines. During my onsite visit, Fox poured three examples of her delicate, subtle pinot noirs. The 2017 Hyland Vineyard Coury Clone is still tight but elegant. The 2018 Maresh Vineyard Star-of-Bethlehem Flower Block has a creamy cherry and spicy nose, with steely notes on the palate. From own-rooted vines planted in 1970, the 2018 Maresh Vineyard offers a captivating mix of earth and delicate floral notes.

Tahmiene Momtazi and her father Moe of Maysara Winery treated me to tastes of seven wines at their estate. The three I sub-mitted include the 2019 Arsheen Pinot Gris, a portion of which was aged in a concrete egg, resulting in great texture and mouthfeel. The explosive 2016 Jamsheed Pinot Noir, always an outstanding value, and the 100-percent whole-cluster-fermented 2015 Rumi Pinot Noir are also in.

Chris Berg, of Roots Wine Co., and I tasted over Zoom. He had given me two pinot noirs and two whites. The beautifully balanced 2019 Pinot Gris is redolent of grapefruit. The 2019 Sauvignon Blanc is barrel-fermented, which yields a smooth mouthfeel to complement the great acidity. Both are excellent values.

Having received a particularly abundant crop from his source, Berg has been experimenting with various expressions of sauvignon. Exhibiting dark fruit, flowers, and polished oak aromas, the 2018

Saffron Field Pinot Noir is surprisingly delicate on the palate, with a long finish, fine tannins, and acidity.

Because the tasting room had not opened, Jason Lett, with mask in place, dispensed seven classic wines from The Eyrie Vineyards from his Coravin, making it particularly difficult to choose which to highlight. The 2018 Estate Pinot Gris offers a rich complex aroma of pears and flowers, with mouthwatering flavors, beautiful acidity, and salinity. Sourced from the fifty-seven-year-old vines, the 2017 Original Vines Chardonnay is an elegant and delicately savory example, with just a hint of oak and great balance. While drinking well now, the 2016 Outcrop Pinot Noir has many more years to go. The nose is juicy and fruity with some herbal undertones.

Field coordinators are asked to nominate wineries for awards to recognize notable interpretations of the values of the guide and value for the money and wines for outstanding sensory quality, finest bottles from a sensory point of view, and excellent value.

As Parker Wong reports, "When it's business as usual, everyone gathers for the awards-tasting of all the wines that have been nominated by the field coordinators. We had several hundred last year for California and tasted for two full days. As we will not be meeting this year, awards will be given based on the nominations by the field coordinators and approval of the directors…"

The results were included in the *Slow Wine Guide 2021*, which was released early that year.

During this extraordinarily unhurried period, "slow wine" took on a meaning beyond that relating to its production and origin. With my schedule more open, I was able to linger over the wines, letting them evolve over time. This was especially important since most were young and some were recently bottled. The overall excellence and variety that I tasted unquestionably ensure that Oregon's wine reputation will remain outstanding in the field.

A Quick Look at Slow Wine

Visits to wineries selected for inclusion in the *Slow Wine Guide USA 2024* wrapped up at the end of August 2023. The guide, an offshoot of the Slow Food Movement that started in Italy, only includes "good, clean, and fair" wines grown without synthetic herbicides.

Starting in spring, field coordinators, including me, fanned out across California, Oregon, Washington, and New York to taste with winery owners and winemakers and to collect updates to existing listings or to establish new ones. For each winery, an entry is prepared that includes a section called "Life" that tells a bit about the history and people. Then, there are sections describing the vineyards and wines. A farm audit concludes the listing, which summarizes methods used for caring for the soil and vines.

My beat is the Willamette Valley. Here are the highlights of my tour.

Abbott Claim Vineyard is Anthony Beck's project in the Yamhill-Carlton American Viticultural Area (AVA). French winemaker Alban Debeaulieu is producing beautifully textured pinot noir and chardonnay that are approachable young yet structured to age. The 2019 Abbott Claim Vineyard Pinot Noir exhibits a dense complex nose and a delicate, well-balanced palate.

Belle Pente Vineyard & Winery was founded in 1994 by Brian and Jill O'Donnell. Brian creates fresh, age-worthy wines. From the

Murto Vineyard in the Dundee Hills, he produced a youthful nicely balanced pinot noir in 2019.

Josh and Caroline Bergström oversee Bergström Wines, which was founded by Josh's parents in 1999. Chardonnays go through full malolactic fermentation and rest on a solera blend of lees. Pinot noirs are fermented whole-cluster to enhance structure and complexity. The 2021 Sigrid Chardonnay emits complex aromas of toasted hazelnut, flowers, lemon, and with air, graham crackers. The briny palate is light and mouthwatering, with a lemony finish.

Björnson Vineyard was founded by Pattie and Mark Björnson in 2006. Pattie serves as executive winemaker and turns out wines made from an increasingly wide selection of varieties that display bright acidity and intense flavors. The 2021 Magnus Pinot Noir has a rich, handsome nose and an assertive, yet graceful palate.

Doug Tunnell, founder and winemaker of Brick House Wines, fashions wines at his Demeter-certified biodynamic estate in Ribbon Ridge AVA. His 2021 Les Dijonnais Pinot Noir, fermented with 30-percent whole clusters, exudes an intense floral nose and has a rich palate.

When Jon and Kathy Lauer acquired Bryn Mawr Vineyards, they engaged Rachel Rose as winemaker and vineyard director. Rose crafts wines that highlight the aromatics resulting from the winds through the Van Duzer Corridor. Her 2019 Estate Pinot Noir is intensely aromatic, with a retronasally perfumed palate and smooth tannins.

Chris and Melissa Thomas purchased a twenty-year-old vineyard in 2020 and renamed it Celestial Hill Vineyard. They produce pinot noir and chardonnay with bright acidity and compelling aromatics. With an aroma reminiscent of line-dried laundry, the mouthwatering, saline 2021 Yamhill-Carlton Chardonnay begs for food.

New to the guide in 2024 is CHO Wines, founded in 2020 by Dave and Lois Cho. Dave makes acid-driven wines with lower alcohol, achieving delicacy over power. From the Tualatin Hills AVA, the 2022 Pinot Noir Pétillant-Naturel is a fun, refreshing bubbly offering bright cherry and watermelon on the nose and palate.

Corollary Wines, founded by Jeanne Feldkamp and Dan Diephouse, is dedicated exclusively to making Willamette Valley sparkling wine. Utilizing the traditional method and extended aging on lees, they produce beautifully textured and richly complex bubbles. The 2018 Winter's Hill Pinot Blanc has aromas of brioche and apple, with an intense lemony palate, great texture, and long finish.

Cramoisi Vineyard was planted in 2012 by Sofía Torres McKay and her husband, Ryan McKay. From it, more powerful complex expressions of pinot noir and rich, well-balanced chardonnay are made. The 2019 Dundee Hills Cuvée Pinot Noir is intensely floral on the nose, with a harmonious palate featuring red cherries and good acidity.

Venerable David Hill Vineyards & Winery in the Tualatin Hills AVA is the home of some of the oldest vineyards in the Willamette Valley planted by Charles Coury. Chad Stock fashions cerebral yet elegant whites emphasizing finesse over overt fruitiness. Old vine pinot noirs have complex aromas and great texture. The 2021 Whole Cluster Pinot Noir exhibits spice, bright fruit, and wood on the nose, with a young palate, smooth tannins, and excellent texture.

The venture of husband and wife Patrick Reuter and Leigh Bartholomew, Dominio IV, produces a range of varieties using a minimalist approach to preserve acidity, aromatics, and ageability. The nose of the 2012 Voyager's Manuscript Tempranillo, made from grapes from the Columbia Gorge, has great depth, evoking dark fruit in a wooden frame. The understated silky palate is of uncommon elegance with a medium-long finish.

Marcus Goodfellow and his wife Megan handle all aspects of production of Goodfellow Family Cellars. They create structured pinot noirs that are typically 75-to-100-percent whole-cluster fermented and puncheon-aged chardonnays. A 2019 Lewman Vineyard Pinot Noir Heritage No. 16 is beautifully perfumed with spice, flowers, fruit, and stem influence. The youthful palate displays great acidity and structure.

Jay Somers and his wife Ronda Newell-Somers own J. C. Somers Vintner, which fashions wines from grapes sourced from several vineyards. The 2021 La Colina Vineyard Pinot Noir offers a pretty, cinnamon-spice, bright-cherry nose, and a young but serious dark-cherry palate, with a medium finish, good acidity, and fine balance.

Another first-timer is Lafayette and White Cellars, a small-batch producer started by Danielle Lafayette and Andrew White. They employ minimal intervention and novel approaches to both expected and less common varieties. The 2022 Embue Pinot Gris is an amber wine sourced from the Tualatin Hills AVA. It offers aromas of fruit and spice with savory notes and yields a nice perfume of cantaloupe as it warms. The palate features sweet melon.

Maysara Winery is a Momtazi family affair, with founders Moe and Flora supported by daughters Tahmiene, who makes the wine, and Naseem and Hannah who assist with sales and events. Momtazi vineyard in the McMinnville AVA supplies all of the fruit for beautifully balanced wines made with no additives. The stunning 2007 Mitra Pinot Noir is a classic example of cool-vintage, cool-climate pinot. The nose highlights dark fruit and mint over savory notes. The palate is after the nose, elegant with medium-finish and fine tannins.

Nicolas Jay is the joint project of Burgundian winemaker Jean-Nicolas Méo and music entrepreneur Jay Boberg. Together with associate winemaker Tracy Kendall, they produce elegant pinot noir and chardonnay that are complex, structured, and capable of aging. The gorgeous 2021 Affinités Chardonnay offers aromas of juicy fruit, hazelnuts, and the impression of chalk. Fruit plays across the lovely delicate palate, which displays freshness, great finesse, a hint of butterscotch, and a medium finish.

Remy Drabkin founded Remy Wines in 2006 and serves as the winemaker. She makes Old World-style wines from a range of varieties, specializing in those from Italy, with minimal intervention and thoughtful use of new oak. The nose of the 2019 Dolcetto, from Jubilee Vineyard in the Eola-Amity Hills, bursts with fruit and some new oak. The elegant juicy palate offers excellent acidity, a medium finish, and plenty of structure for ageability.

Industry veteran Harry Peterson-Nedry started Ribbon Ridge Winery as a side project in 2002. His daughter Wynne Peterson-Nedry joined in 2009 and is now the winemaker and proprietor. The wines are characterized by judicious use of new oak and occasional whole-cluster fermentation for the pinot noir and bright acidity for the whites. The 2022 Ridgecrest Grüner Veltliner exudes bright aromas of white flowers. The mouthwatering palate is round and fruity, with great balance and a medium-long finish.

Chris and Hilary Berg started Roots Wine Co. in 2001. Chris specializes in crisp, sometimes weightless, bottlings of whites and complex aromatic reds plus uncommon blends and treatments. From the Estate East Vineyard in the Yamhill-Carlton AVA, the 2022 Rosé is a pale pink made from trousseau, sauvignon blanc, and pinot noir and yields mixed fruit aromas and a rich, juicy palate.

Since 2005, Jason Lett has headed The Eyrie Vineyards founded by his late father, David, and his mother, Diana, in 1965. His emphasis is on ethereally elegant wines with excellent structure to ensure longevity. Though still young, the 2017 The Eyrie South Block Reserve Pinot Noir from the original plantings in the Dundee Hills is intensely perfumed and pretty as can be, with a seemingly weightless palate. Some tannin is apparent in this nascent beauty.

Winderlea Vineyard & Winery was founded in 2006 by husband and wife Bill Sweat and Donna Morris. Sweat makes elegant wines with consultant Robert Brittan, emphasizing delicacy and complexity, with refreshing acidity and fine tannins to encourage ageability. From the McMinnville AVA, the deep yellow 2021 Meredith Mitchell Pinot Blanc has a nutty nose and rich, lingering palate, with great balance and long finish. It makes one wonder why more of this variety isn't planted.

Deborah Parker Wong, the US national editor of the guide, and Pamela Strayer, the US senior editor, combed through the draft entries and reviewed and approved the nominations made by the field coordinators for wine and winery awards. They produced the 2024 guide, which was released in December 2023. For more information, go to https://slowfoodusa.org/product/slow-wine-guide-usa-2024/.

PART VIII

Grape Explications

Grape Explications

Musings of a Novice Wine Writer

Admittedly, it is presumptuous of a novice to be writing about writing about wine. This is especially true of one who had no particular vision when he started. Since I have been moving away from inchoateness, the fifth anniversary of my first popular wine article is a good milestone to reflect on how I got to do what I never planned to.

While numbers first attracted my interest and became the focus of my education and career, words have always been an irreplaceable source of pleasure. I developed literacy in parallel with numeracy. At the same time I was immersed in math, English teachers in high school and during my freshman year in college firmly, sometimes mercilessly, shaped my writing. Liberal arts courses frequently required essays, affording me the chance to hone what I had learned. With one exception, the only writing I did in graduate school was my dissertation, in which the ratio of words to symbols is only slightly higher than the odds of finding a revelatory bottle of wine in the "under $5" section.

That exception was a collection of tasting notes I published, in a few issues of the long-defunct *Vintage* magazine in the 1970s, under a byline that included the parodistically pretentious title of Minister of Wine, Duncan Hines Memorial Bon Vivant Fellowship, Int'l. (see p. 10) These were compilations of members' reactions to some truly wonderful French and German bottlings. I continued to fill notebooks with details about every wine I tasted through graduate school, which amuses dinner guests to this day. (see p. 9) My writing after graduation, however, was exclusively technical and definitely non-vinous.

This changed, in 2009, when I discovered a reason to write a single piece that included both math and wine. I had become fascinated with the mathematics behind aggregating preferences. The movie *Bottle Shock* raised my curiosity about how the judges arrived at their rankings and concluded that California wines topped the red and white lists. After more digging deeper than I expected to have to, I prepared "The Judgment of Paris According to Borda."

It details the flaws in the system used and applies a mathematically defensible method to show that, while a California chardonnay did, in fact, belong in first place, a French wine took top honors in the red category. It is the first article in an issue of the *Journal of Wine Research* that same year.

Although the analysis in it isn't particularly esoteric, the article appealed exclusively to those not plagued by math anxiety. Ironically, it was of interest because of the accessibility of the description of the recommended method rather than for its application to wine tasting and for its contradicting the outcome of the celebrated event.

Around the time the article appeared, we bought our house in McMinnville so that we could enjoy life in a fabulous wine region without having to learn a new language or currency. After we became full-time residents of Oregon in 2011, I wrote my first popular article, "Borda is Better," (see p. 16) that explains the method to the lay reader. It appeared in the October issue of the *Oregon Wine Press* that same year and was my first for that publication.

So now, I was hooked. As an outlet for my more rigorous inclinations, I joined the American Association of Wine Economists and regularly present at meetings and contribute to its *Journal of Wine Economics*. Along with the *Journal of Wine Research*, this periodical has published my reviews of wine-related books, allowing me to keep up with the work of more established wine writers.

On the popular front, I occasionally don a press hat to attend and write about a wide range of wine-related events. I am deeply flattered that invitations to these have increased recently as the gatherings are certainly entertaining and give me access to some memorable vintages. But my preference is to write about things that amuse me even if the potential appreciative readership is small. While I still collect a healthy number of rejections, I have been fortunate to have articles appear in a few important publications.

Generally, I no longer keep detailed notes since formulating and recording them tends to distract me when I'm tasting. Enjoying wine, for me, is a wholly right-brained activity while writing exercises the other hemisphere. Although I certainly utilize the latter to prepare articles and do math, since I don't do analytic tasting, I feel no compunction in not mixing the two sides. I want either wine or words and numbers in my mouth and mind, but not both at the same time.

I will point out notable wines from events I attend, but I won't score them. As a card-carrying mathematician, I regard doing so as felonious number abuse. Many have railed against the absurdity of reducing the experience of enjoying one of nature's and man's finest collaborations to a single number.

Yet the persistence of this practice strikes me as ironic in a culture that prides itself on its innumeracy. I hear all too regularly: "Oh, I've never been good at math." Yet no one would ever brag about being illiterate. At the same time as boasting about the inability to comprehend even the most basic mathematics, many of my countrymen imbue numerical ratings with importance well beyond anything conveyed by their content.

I was recently asked if I had any special training that credentialed me to write about wine. I do not. My only qualifications are that I have a respectable list of technical, trade, and popular publications and have been sampling the fruit of the vine in five continents for nearly half of a century. Unlike the apprenticeship to a mentor I went through in graduate school to earn a doctorate in applied mathematics, I fill in the gaps in my background by working at a winery, listening to the professionals, and especially reading constantly.

There are several writers I admire. Alex Hunt MW, a prolific and excellent essayist in his own right, praises three—Andrew Jefford, Hugh Johnson, and Terry Theise—by benevolently regretting his inability to communicate the emotions elicited by a remarkable Italian wine with "Jeffordian poetry, Johnsonian prose, or Theisian powers of self-analysis and self-revelation." (http://www.worldoffinewine.com/news/soldera-the-great-outsider-4762352/)

I stumbled into the increasingly crowded field of wine writing five years ago, with no particular objectives in mind except to entertain myself and whomever might enjoy what I have to say. As I continue writing about this wonderful beverage, I hope that what I produce mirrors what I value most in a memorable wine: elegance, the virtue most sought after by mathematicians; complexity; and finesse. A lingering finish never hurts, either.

When it first appeared, this piece on spitting was very popular among my colleagues and visitors to the tasting room where I work. It made quite a splash.

Grape Expectorations

Confessions of a Novice Spitter

An arc of Beaune Clos des Mouches Premier Cru Rouge sails from the lips of Jean-Pierre Cropsal to a narrow opening in a nearby receptacle. Confidently, it disappears silently without leaving a drop anywhere. If there was a Wine Spitters' Hall of Fame, Cropsal would have to be among the charter inductees. I had only heard about his uncommon aim prior to my visit to Burgundy in June 2013 where Cropsal was our guide through the cellars of Joseph Drouhin. There, beneath the streets of Beaune, as we sampled six bottles, I witnessed his mastery of expectoration with a combination of awe and envy.

Until recently, I had never spit wine out. I never really had to nor really ever wanted to. In most cases, I was drinking with meals or at least with some food. Given the rarity of many of the wines, it seemed like a terrible waste not to swallow, or *cracher à l'interieuras* ("to spit on the inside"), as I learned it is called in France.

All this changed when I started to cover wine events and work in a tasting room. Early on at tasting events, where I needed to record impressions for articles, I would swallow a bit, fearing that I would lose too much of the aftertaste, then dump the rest. It wasn't until I started working at the winery, where it was forbidden to drink while on duty, that I needed to spit out the wine I sampled when opening a new bottle.

This was problematic. If, as it was noted during World War II, that loose lips sink ships, I quickly discovered that they could also result in wine-stained white shirts. Like the earth before creation, what I emitted was without form. So now, I just lean over the sink and spray into it. Inelegant, a bit noisy, but effective.

At seated staff tastings, however, my approach is different. I prefer to spit into a glass rather than the personal wine spittoon favored by my employer. The latter has a raised center, presumably intended to prevent backsplash but which I invariably hit straight on, thus defeating its purpose. To avoid problems, I apply no pressure and just let the liquid drop. Happily, my concerns about lost or diminished aftertaste never materialized, and I now spit rather than swallow when performing in my journalistic role.

Oh, the wines I have spit! At blind tastings of pinot noirs held for the winery staff between 2012 and 2014, the stars of the spit bucket list included 2009 DuMOL Finn from California; Oregon's 1999 Cristom Marjorie Vineyard; 2000 Ken Wright Canary Vineyard; and 1999 and 2001 Beaux Frères The Beaux Frères Vineyard.

Burgundy notables included 2009 Gevrey-Chambertin from Méo-Camuzet, 2009 Chambertin Clos de Beze Grand Cru from Bouchard Père et Fils, and 2001 Richebourg Domaine de la Romanee-Conti. The average price of the latter at this writing is $1,648 or about $65 an ounce. Had I failed to hit the spittoon, the stain would have been worth more than the shirt.

In June 2013, I joined Scott Wright on a six-day group tour of Burgundy producers whose wines he imported. Over that period, we tasted almost one hundred different wines, with well over thirty on one of the days, from barrels and bottles.

Usually, I only swallowed at meals and spit in the cellars and tasting rooms in order to remain vertical. Some of the winemakers we visited were Domaine Thibert in Fuissé, Huber-Verdereau in Volnay, Buisson-Charles in Meursault, A-F Gros & Caroline Parent in Beaune, Marc Roy in Gevrey-Chambertin, and J-J Confuron in Premeaux-Prissey.

In the cellar of Domaine Comte Georges de Vogüé in Chambolle-Musigny, winemaker François Millet extracted barrel samples from some truly remarkable premier and grand crus from the tiny 2012 vintage. Only three barrels of the Musigny Vieilles Vignes were made, yet we did get a small taste that I could not spit out.

In 2013 and 2014, I was invited to cover Passport to Pinot, the condensed version of the International Pinot Noir Celebration (IPNC), and the full affair in 2015. Both took place at Linfield College in McMinnville, Oregon, about one mile from my house. The shorter event is held outside in the oak grove.

After nearly thirty years, the soil is well nourished by the effluence of thousands of tasters, including me. Failure to spit would have necessitated my corralling a designated walker to lead me home. For the full IPNC tastings conducted indoors, I made good use of the plastic cups provided. Wines of note that passed my lips in both directions at these events are called out at Passport to Pinot: Something to Walkabout (p. 151), IPNC for the Rest of Us (p. 155), and A Random Walk through IPNC (p. 159).

Becoming a believer, if not a skilled practitioner, in spitting during serious tastings puts me in very good company. In *The Taste of Wine*, Émile Peynaud observes: "When at work professionally, tasters generally spit out as much wine as possible. Not because this improves the act of tasting, the reverse if anything [is true]; but during the course of tasting anything from ten to thirty wines…it would clearly be impossible to swallow several mouthfuls of each…without some ill effects."

Spitting, "an essential practice at professional tastings," merits its own entry in *The Oxford Companion to Wine*. Perhaps tongue in cheek, Jancis Robinson notes that: "Members of the wine trade, and wine writers, rapidly lose any inhibitions about spitting in public."

Ironically, this may not be the case for Great Expectorator Jean-Pierre Cropsal. Domaine Drouhin had no photograph of him in action despite his legendary prowess. It was even speculated that they were "not sure he will be very happy to be pictured in such an elegant position."

I do wish that more folks who visit tasting rooms serially would learn from the professionals. Especially toward the end of the day, I have faced many of the premoistened who really ought to expect to expectorate to avoid becoming overmoistened. They should save swallowing for mealtime.

My prowess at spitting is closer to Lewis and Julianna, my vintage 2015 grandchildren, than to Cropsal. And while they will someday outgrow dribbling, I'm not so sure I ever will. For me, world-class spitting will forever remain an expectator sport.

Happily, Lewis and Julianna, along with Liam (vintage 2013), Myriam (vintage 2016), Ethan (vintage 2017), and Shayna (vintage 2019), are all past the dribbling age. Teddy (vintage 2021) is also past this stage but was joined in 2024 by his brother, Murphy, who is just starting.

The Value of Elegance

Even though it lacks a formal definition, like so many other popular wine descriptors, "elegant" has been appearing in tasting notes for ages to describe a wine that is the opposite of one that is bigger, bolder, simpler, and rustic. Despite the lack of clarity, this term became a regular part of my vocabulary during the 1970s, when tasting great wines for the first time while also studying and practicing mathematics.

In math, the proof of elegance is in the proof, whereas with wine, the evidence of elegance is more elusive. The wine blogger and author of *The Wine Bible*, Karen MacNeil, concluded: "In wine, elegance isn't so much about certain flavors—as much as how those flavors present themselves. …In the end, elegance has a quiet beauty. Wines with exquisite elegance compel you to drink them—not because they leave you impressed, but because they leave you vulnerable to a deeper kind of contentment." (https://winespeed.com/blog/2017/05/what-is-elegance-in-wine/)

As graduate students, we were taught some exquisitely elegant proofs, the impact of which was not only intellectually stimulating but aesthetically satisfying. Ray Li, an assistant professor in the Mathematics and Computer Science Department of Santa Clara University, offered three criteria for identifying them: "An elegant proof is unexpectedly simple[,]…hits at the heart of the problem's complexity[, and]…reveals more about the subject than simply its result." (https://www.quora.com/What-makes-a-mathematical-proof-elegant#:~:text=An%20elegant%20proof%20is%20a%20simple%2C%20unexpected%20and%20coherent%20proof) This characterization not only defines an elegant proof but also highlights its value.

How nice it would have been if the proofs in my dissertation could have earned that level of praise, the highest a mathematician can bestow. But alas, while they served the purpose of establishing that which needed to be demonstrated, they are more appropriately described as functional, computational, and brute force. One can easily see the similarity between such a proof and a finesse-less "big-ass" red wine that hits you over the head with fruit and oak.

Can an encounter with an elegant proof lend insight into the experience of tasting an elegant wine and the value of doing so? I think so, but not in an obvious way. I will not attempt to distill a conclusive definition of the word or offer a synthesis of the two presented above. Instead, I will explain the effect of encountering elegance that I suggest creates its value.

Decades after the first time I used the term, it was a meal, not a particular wine—though sake was involved—or a mathematical proof, that crystalized the importance of elegance for me.

To celebrate the promotion of my elder son to an executive position, the two of us went to dinner in October 2021 at Sushi Ginza Onodera in West Hollywood, California. For just under two hours, we and three young couples sat in near silence around the unadorned wooden bar in this sliver of a restaurant as an array of exquisite morsels was sliced, plated, and placed in front of us. From the understated décor to the server who unobtrusively delivered ornate, oversized platters, on which rested the strange and the familiar and poured Shichida "Junmai Daiginjo" into small, colored, cut glasses, I was immersed in elegance. More to the point, I was focused for the entire time on what was transpiring and enjoying my son's company. I was entirely present.

The meal cost $400 per person, with the bottle of sake adding another $230, so it was a "grand" experience. Yet as I thought about it on the long walk back, perhaps in the afterglow of "being in the moment," I concluded it was money well spent. Being enveloped in elegance had eased me into mindfulness.

When one is going through the daily routine, attaining mindfulness isn't usually on one's mind. Even now, freed from the obligations of a full-time job and raising a family, with little on my daily schedule, I still follow a regular routine that doesn't require much attentiveness. But still, with most of my future behind me, how much more important is it to pay closer attention to what is going on?

An elegant proof grabs your attention. As Li explains, such a proof, though simple, gives insight into the intricacies of the matter that also might be present elsewhere. Four years after I completed my dissertation, my advisor and I published the generalization of its results. Though the proofs are simpler, they aren't necessarily simple, and they do reveal insights into the complexities of the situation. They also recast the problem in a manner that not only bypasses a limitation in the method I used previously but also suggests a framework for approaching a problem in an entirely different area. The proofs don't rely on brute-force computations. If not outright elegant, this work is certainly less inelegant than my dissertation.

After turning twenty-one and thus able to buy my own alcohol, I found wine an increasingly fascinating diversion when I was grappling with my research. In reviewing the four hundred and fifty tasting notes I compiled between 1969 and 1979, the period that included my time in graduate school, I found thirty that contained "elegant."

For example, about the 1964 Petrus tasted on March 12, 1977, I highlight its "[e]legant, very dark brick with orange edges," then go on to rhapsodize about its "[l]ovely perfume" and "rich complex taste" before praising, oxymoronically, its "[l]ong-lasting finish...both subtly elegant and bold." (All this was had for $20, by the way. On March 25, 2022, wine-searcher.com listed the average price as $4,321.) I used the term to describe not only the wine's finish but also the color. The latter was the case for many of the three dozen notes, not surprisingly, since the eyes drink first. While my memories of almost all of that period are long faded, I suspect that my attention was captured by the wine's appearance and then sustained by its nose and palate. The wine consumed me while I consumed it.

In addition to the Petrus, some of these old wines are still available, in case you wish to see how well they have held up, but it will cost you. For the other eleven of the thirty-six "elegant" wines that wine-searcher.com listed as currently or recently available, here are the original prices paid, followed by the average price on March 25, 2022, for a 750 ml bottle, unless otherwise noted: 1968 Heitz Napa Valley Cabernet Sauvignon ($16 in 1976, $494), 1959 Château Beychevelle ($18.14 in 1977, $457), 1964 Château Brane-Cantenac ($10 in 1976, $174), 1966 Château Brane-Cantenac ($10.49 in 1977, $178), 1970 Château La Gaffeliere ($4.29 for 375 ml in 1976, $150), 1964 Château Leoville Barton ($7.98 in 1974, $276), 1959 Château Leoville Las Cases ($15 in 1974, $573), 1971 Château Rieussec ($3.79 for 375 ml in 1975, $203), 1970 Gevrey Chambertin Armand Rousseau ($7.39 in 1975, $636 in 8/20), 1972 Morey Saint-Denis Domaine Dujac ($10 in 1977, $539 in 5/20), and 1947 Musigny Ponnelle ($6 for 375 ml in 1970, $1,428).

If these are too rich for you as they are for me, or if you are concerned about their elegance aging out, as I would be, I recommend alternatives from a much more recent vintage noted for its elegance and, in many cases, still available. Here are six "elegant" 2011 Willamette Valley pinot noirs that I tasted in early 2021 as part of my ongoing exploration of wines made in the last truly cool year in Oregon, with prices I paid in 2013, except as noted, including discounts for a 750 ml bottle followed by the current price at the winery: Domaine Drouhin Oregon Dundee Hills ($34, $90) and Laurene ($55.25, $130), Saffron Fields Yamhill-Carlton District ($36, $80), RR Ridgecrest Vineyards Ribbon Ridge ($62.50 in 2021, $89), and Chehalem Statement Ridgecrest Vineyards ($69, $140).

There is an important distinction between tasting and drinking wine. The former is done intently; the latter is not. Wine writer and author Terry Theise reminds us: "Usually the way a wine tastes is indeed different from the way it drinks when we're not paying obsessive attention to it." (https://worldoffinewine.com/2022/02/10/mousse-fils-meunier-of-another-type-entirely/)

Since I prefer to have wine with food, I take time to sip it by itself at various times during the meal so that there is no distraction from becoming more deeply engaged.

While tasting a wine of elegance isn't necessarily the only way to trigger mindfulness, it is sufficient, to use a common mathematical expression. A fruit-forward wine can grab your attention because it is obvious, but an elegant wine seduces you, drawing you in as it reveals its subtleties, and like an elegant proof, its complexities. By inducing mindfulness, both create a more profound awareness of what wine and mathematics are.

For me, the word "elegant" now has a meaning beyond just another example of winespeak. It is an indication that the beverage should command my complete concentration in much the same way an elegant proof would. I challenge a "neuroenologist" to see if there is a different reaction in the brain when consuming a wine deemed elegant compared to one that is not.

Instead of taking up residence in the metaverse, interact with a glass that contains an elegant wine rather than a screen full of intangible images. It will bring you back down to this earth. Seek elegance and when you find it, you'll find presence. And how priceless that would be even for a brief moment.

Vindicating Thomas: Virginia Vanquishes Vinifera

In the end, there was no reason to doubt Thomas—Thomas Jefferson, that is. *Vitis vinifera* now flourishes in the Old Dominion. In 2010, the Commonwealth of Virginia ranked eighth in the United States in total grape production and in bearing acreage. The national and international reputation of wines from the state is rising as evidenced by numerous accolades in prestigious publications and competitions. It is not unreasonable to believe that Virginia will soon be able to claim the title of the best American wine-producing state east of the Rocky Mountains. But the impending success didn't happen overnight; it was over four hundred years in the making.

Though settlers in Jamestown produced wine from English grapes in 1609, the commercial wine industry in the US is generally considered to have begun in 1619 with the passage of "Acte 12" by the General Assembly of Virginia, also in Jamestown.

It decreed "that every householder doe yearly plante and maintaine ten vines, until they have attained to the arte and experience of dressing a Vineyard, either by their owne industry, or by the Instruction of some Vigneron. And that upon what penalty soever the Governour and Counsell of estate shall thinke fit to impose upon the neglecters of this acte."

The intent was to make Virginia a significant source of *vinifera* wines for the British Empire. Unfortunately, nature conspired against the plan by inflicting diseases and pestilence upon the vines. Then, as today, the climate wasn't too cooperative either. Soon the vines were replaced by tobacco, which proved a much quicker path to wealth.

But, as economic cycles go, by 1759 the tobacco trade was in a depression. Charles Carter chaired a committee of the Virginia Assembly that was formed to look at diversification. He and his brother Landon had already been growing and vinifying both native and *vinifera* wine grapes, so commercial winemaking was high on his list of proposals. The Society for the Encouragement of the Arts, Manufacture, and Commerce (now known as the Royal Society of Arts) in London encouraged and advised Carter in his vinous pursuits.

In 1762, the society awarded him a gold medal in acknowledgment of his "spirited attempt towards the accomplishment of their views, respecting wine in America" on the strength of two samples—one made from an American winter grape and a second, from a white Portugal summer grape.

One year later, Colony of Virginia Royal Governor Francis Fauquier certified Carter's success in growing both red and white European grape varieties. In 2012, the Virginia Senate passed a resolution to commemorate "the 250th anniversary of the first internationally recognized fine wines produced in the Commonwealth as a result of Charles Carter's award from the Society for the Encouragement of the Arts, Manufactures, and Commerce."

Colonel Robert Bolling Jr. was another advocate of *vinifera*, having found the wines made from native American grapes too acidic. Between 1773 and 1774, he produced a manuscript entitled *A Sketch of Vine Culture for Pennsylvania, Maryland, Virginia, and the Carolinas*. It contains collected wisdom from a wide range of European sources and also specific recommendations, based on his own experience planting and tending a vineyard in Chellow, in the central Virginia county of Buckingham.

In addition, Bolling's "Essay on the Utility of Vine Planting in Virginia," the lead article in the February 25, 1773 issue of *The Virginia Gazette*, exhorts the House of Burgesses to fund public vineyards and experimentation thereon with imported grape varieties: "Let us, with utmost Expedition, provide Vineyards in various remote Counties, and in Places where it would be lost Labour to cultivate Tobacco." Bolling was granted £50 annually for five years by the House. Unfortunately, his experiments were cut short by his death in 1775.

It isn't known if Thomas Jefferson had knowledge of the success of Charles Carter or of the works of Bolling, though the two were well acquainted and shared relatives when he planted grapevines at Monticello in 1771, thus beginning his ill-fated experiments.

Two years later, Jefferson's friend Filippo Mazzei arrived from Italy to assist at both Monticello and Colle, a nearby farm. The reason these efforts failed to yield a crop is a matter of debate. While phylloxera, a fungal disease, and inclement weather have been blamed, Gabriele Rausse, Monticello's current assistant director of gardens and grounds, theorizes that birds and/or deer were responsible.

In response to the lack of success with *vinifera*, hybrid grapes—crosses between two or more *Vitis* species—took center stage in Virginia during the nineteenth century. One in particular, Norton, first cultivated in the early 1820s by Dr. Daniel Norborne Norton in Richmond, Virginia, remains popular.

Norton is believed to be descended from *Vitis vinifera*, possibly enfariné noir, and *Vitis aestivalis* and is identical to cynthiana. By 1830, the grape was marketed commercially by William Prince of Flushing, New York, who received it from Dr. Norton. Norton/cynthiana was also successfully planted in Missouri, where it was named the State Grape in 2003, as well as in other states east of the Rockies.

In 1873, the Monticello Wine Company was founded in Charlottesville, and in the same year, its Norton-based "Extra Virginia Claret" earned national and international acclaim. Today it is used to make a range of styles, from rosé to dry red table wine to port. Leading Virginia producers include Horton Vineyards and Crysalis Vineyards.

Monticello Wine Company became the largest winery in the South, but its success was cut short on November 1, 1916, when Prohibition infested the Old Dominion more than three years before the 18th Amendment took effect.

In 1949, sixteen years after the 21st Amendment restored some sanity, John Baptista Sciutto started his eponymous winery on the grounds of the Manassas Battlefield Park. Until he was forced to close in 1955, he produced wine from Norton and *vinifera* grapes.

In his book *Beyond Jefferson's Vines: The Evolution of Quality Wine in Virginia*, Richard G. Leahy writes: "...the renaissance of the Virginia wine industry began in the late 1960s around the town of Middleburg, in the northern Piedmont east of the Blue Ridge Mountains...The post-Prohibition pioneers grew hardy French hybrid varieties...to avoid the vineyard woes experienced by Jefferson and the Jamestown settlers, setting the pattern for small Virginia wineries for the next twenty years."

In 1973, the Vinifera Wine Growers Association (now the Atlantic Seaboard Wine Association) was founded and, in the same year, planted chardonnay in the Piedmont Vineyards, making it the first commercial *vinifera* variety after Prohibition in Virginia. Favorable legislation followed in the 1980s that created a most congenial environment for growing and producing wine from Virginia grapes. In 1984, a fund was created using an excise tax on Virginia wine to subsidize research and staff positions at Virginia Polytechnic Institute and State University (aka Virginia Tech), which has since become a major contributor to the growth and success of the state's wine industry.

Also around this time, the emphasis shifted from the French hybrids to *vinifera*. Sometime in the 1990s, I remember tasting my first Virginia wine, a completely forgettable riesling, at an American bistro in Reston, which at the time only carried domestic bottles, of which this was the sole representative from the state. Along with cabernet sauvignon, riesling plantings diminished because of its inability to handle the humidity. The production of *vinifera* grapes finally overtook French hybrids in the mid-1990s.

While it may no longer be true in biogenetics that ontogeny recapitulates phylogeny, the fact that the theory of recapitulation appears to hold for wine regions is illustrated by the evolution of the Virginia wine industry. Viticultural areas around the world go through a decades-long—historically, centuries-long—process of identifying the grape varieties suitable for a site's microclimate and soil.

Virginia is finding success with viognier, cabernet franc, and petite verdot. The industry is typically aided by academic institutions that train future generations of wine-grape growers and winemakers and conduct innovative research into farming and winemaking methods.

Virginia Tech fills that role in the Old Dominion. Sensible legislation enables responsible growth. Leahy notes: "Important steps forward were realized in 1980 and 1985, with the passage and reauthorization of the Virginia Farm Winery Act, establishing favorable tax and regulatory status for wineries with their own vineyards and small commercial production using Virginia grapes."

Consultants and itinerant winemakers are employed to bring state-of-the-art practices home to the point of production. Virginian Lucie Morton shares her considerable expertise with several Commonwealth wineries. Claude Thibaut moved to Virginia from Champagne to produce lovely sparkling wines. Boxwood Estate Winery employs Stephane Derenoncourt to bring Bordelaise know-how to Middleburg. Savvy marketers like Christine Iezzi of Country Vintner seek to brand wines from newer regions and define their place on the increasingly crowded wine wall.

Since 1983, the Virginia Wineries Association has been acting "on the behalf of the wine industry. [It] promotes viticulture and vintner practices and provides a multitude of wine resources and benefits to its members. [It] holds the Governor's Cup® Competition each year and hosts tastings of the Governor's Cup® Case, twelve of the top-scoring wines of the competition…each year." It has also taken on a more visible advocacy role at the state legislature.

The wine press started noticing Virginia wines and heaping kudos. The prestigious UK wine publication, *Decanter*, and other publications and noted critics regularly give high ratings to many of the Commonwealth's wines. With 210 wineries in 2012, the wine scene in Old Dominion was certainly well along the evolutionary path.

When benchmarking the state's progress, some Virginia winemakers and experts cite Oregon. According to Leahy, Rutger de Vink, founder of RdV Vineyards, a new high-end winery dedicated to Bordeaux blends, "believes Virginia is in a position similar to where Oregon was twenty-five years ago."

According to de Vink, "The challenge in Virginia is that it's too easy for us to sell our wine here. Oregon was forced to improve quality as well as specialize and associate itself with Pinot noir. Now they're planting on the correct soils. My hope for Virginia is that it follows the same trend, going for a focus on site selection, not on what's best for tourism but for where the best grapes are grown."

Leahy writes that Kevin Zraly, author of the *Windows of the World Complete Wine Course*, "feels that Oregon is a good model for Virginia to follow. 'Oregon's wine industry is based on small family wineries, which is also the case in Virginia.'"

From late 2004 to mid-2011, I resided in Alexandria, Virginia. Being both a committed "locapour" and one who lets facts get in the way of opinions, I began to visit nearby wineries hoping to find a more pleasant experience than that unfortunate riesling.

Bordeaux blends, chardonnay, and a seyval blanc from Linden Vineyards were the first to please me. A bargain 100-percent cabernet franc named Marquis de Lafayette from Breaux Vineyards so delighted me that it displaced an Oregon pinot noir as my Thanksgiving wine one year. Breaux's merlots aren't too shabby either.

A 2001 Tannat (as in "what's tannat like?") brought to me from Horton Vineyards and tasted at around five years old could have been mistaken for one from Madiran. Horton's viognier became my standard Thanksgiving aperitif when Oregon pinot noir returned as my red of choice, thus allowing me to feature wines from the two states I was splitting my time between.

The first nebiollo harvested at Three Fox Vineyards was amazing. The Williamsburg Winery Gabriel Archer Reserve from the late 1990s, tasted at about nine years of age, could easily pass for a fine classified growth claret. Boxwood Estate Winery also pays proper homage to Bordeaux with blends made from cabernet franc, cabernet sauvignon, merlot, petit verdot, and malbec. North Gate Vineyard's highly acclaimed 2007 Cabernet Franc was actually sold at the farmers market in my community during the single season the winery had a booth. In little over a decade, Virginia wines have gone from mostly mediocre to progressively more meritorious.

My experience confirms that Virginia wines occupy the middle ground between the Old and New but look more toward Europe than the West Coast for inspiration. The overall impression is stately but not stuffy. There are no fruit bombs; instead, fruit plays nicely on the palate with other flavors. Critics compare Virginia viognier to Condrieu and its cabernet franc to those from the Loire. It is no surprise that the British, historically major consumers of French wine, are going gaga over those from the Commonwealth.

Vineyard farmers and winemakers in Virginia face both different and similar challenges as those in Oregon. Because of the particularly humid climate and rainy summers, sustainable practices seem to play a secondary role to aggressively fighting various diseases and pests, though one winery, Annefield Vineyards, is listed as being biodynamic in Leahy's book. On its website, Chateau O'Brien states that its vines "are nurtured through natural viticulture approaches." Like Oregon, Virginia is increasingly attentive to green building practices. Gold- and platinum-level LEED-certified tasting rooms can be found. Both states are struggling to define the proper balance between alternative uses of wineries for events and other forms of tourism and responsible stewardship of precious farmland.

Into their fifth century, Virginia wines have finally earned a genuine claim to gravitas. Post-Prohibition reconstruction of the industry followed the best practices, resulting in a uniquely American, East Coast expression that still pays homage to Europe. "I always say that there has to be tradition around something to make its branding work," Zraly told Leahy. "Virginia has it more than California, thanks to Thomas Jefferson."

An Oregon-Virginia Taste-off

On Sunday, November 4, 2012, a simultaneous bicoastal blind tasting of Oregon and Virginia viogniers and cabernet francs was held. Tasting panels comprised of distinguished professionals from the food-and-beverage industry convened at the offices of the *News-Register* in McMinnville, Oregon, and at The Wine Loft in Short Pump, Virginia.

The judges in Oregon were Hilary Berg, co-owner of Roots Wine Co. and then-editor of the *Oregon Wine Press*; wine and beer wholesaler Earl Cramer-Brown; restauranteur Emily Howard; winemaker Leah Jørgensen; wine writer Kerry Newberry; and chef and sommelier Michael Stiller. The judges in Virginia were importer Bartholomew Broadbent; wine retailer Booth Hardy; wine distributor Christine Iezzi; sommelier Emily McHenry; and Frank Morgan, wine columnist and blogger.

Three viogniers from each state were sampled in the first flight, followed by a second flight of three cabernet francs from each state. The dozen wines selected to compete survived a careful vetting. The Oregon contenders were picked by me after sampling a wide assortment and included three medal-winners, with one that was Best of Class in the 2012 San Francisco Chronicle Wine Competition.

The Virginia challengers, five of which were gold or silver 2012 Governor's Cup medal winners, were chosen by the organizers on the East Coast. Each of the eleven judges—six in Oregon, and five in Virginia—ranked the wines in each flight, with ties permitted. The Borda count was used to arrive at the consensus ranking (see p. 16). Each judge's ranking was converted to Borda scores as follows: the top-ranked wine received 5; the next, 4; and so on down to the sixth ranked wine which received 0. The Borda scores assigned to each wine by each judge were summed and ordered, from the largest to the smallest sum, to arrive at the consensus ranking.

In both flights, a Virginia offering was ranked first.

The Viogniers	Consensus Ranking
2011 King Family Vineyards Viognier (VA)	1
2010 Narmada Viognier (VA)	2
2011 Spangler Viognier (OR)	3
2011 Troon Viognier (OR)	4
2009 Michael Shaps Viognier (VA)	5
2010 Del Rio Viognier (OR)	6

The Cabernet Francs	Consensus Ranking
2009 Sunset Hills Cabernet Franc (VA)	1
2009 Spangler Cabernet Franc (OR)	2
2010 Troon Cabernet Franc Reserve (OR)	3
2010 Jefferson Family Vineyards Cabernet Franc (VA)	4
2007 Cliff Creek Cabernet Franc (OR)	5
2009 Barboursville Cabernet Franc Reserve (VA)	6

While two Virginia wines led the pack, the Borda scores of the top five viogniers ranged from 33 down to 28.5 indicating that the aggregated judges' preferences were very close. The story was much different with the cabernet francs. The top two had Borda scores of 47.5 and 43 while the remaining four ranged from 24 down to 15.

Virginia Does Pinot Noir?!

Okay, so the eleven judges in Oregon and Virginia ranked Old Dominion viogniers first and second place and a VA cabernet franc first over Southern Oregon bottlings. Maybe we shouldn't be surprised since these two varieties are widely touted as the Commonwealth's pride and joy. But do we Oregonians need to worry that there might be some good still pinot noir coming from Ol' Virginny?

Dennis Horton of Horton Vineyards told Leahy: "I don't think Jesus Christ could grow Pinot noir in Virginia; you can do it, but it doesn't taste like Pinot should."

DrinkWhatYouLike.com blogger Frank Morgan conceded: "In past years, I've found many Virginia Pinots to be insipid wannabes, showing no resemblance to the varietal beauty that can be Pinot noir."

But prompted by an enthusiastic review posted in July 2011 by David McIntyre, *Washington Post* wine critic and blogger, I tasted the 2010 Ankida Ridge Pinot Noir, the first release of this tiny winery made from grapes harvested from vines planted in 2008 on 1.5 acres at 1,800 feet on the slopes of the Blue Ridge Mountains.

To make things interesting, I tasted it side by side with an Oregon pinot noir, the 2010 DePonte Cellars Lonesome Rock Ranch, the first made from grapes planted in 2007 in the foothills of the Coast Range in Yamhill County in the Willamette Valley.

Immediately upon opening, the medium-dark-colored Ankida Ridge put forth an intense, almost jammy, floral fruity nose but sat lightly on the palate with medium tannins. The jamminess did become more pronounced with air. The medium-light-hued Lonesome Rock Ranch offered a more masculine nose of forest floor, fruit, and flowers, with hints of hay and herbs and much tighter structure. Flavors leaned more toward the floral and fruity and, with air, some vanilla and spice. The flavors were more focused; the balance, better; the structure, tighter; and the finish, longer than the Ankida Ridge. This was unmistakably an Oregon pinot noir.

After about three hours of air, a pleasant hint of minerality emerged on the nose of the Ankida Ridge, and the jamminess gave way to very pretty spice over fruit and flowers.

The three who tasted with me definitely enjoyed this effort. Noted wine writers Jancis Robinson and Eric Asimov told McIntyre: "It tasted like Pinot!"

I certainly agree. On the other hand, I also agree with Morgan who wrote in August 2011: "I'm not sure I will ever be completely sold on Virginia Pinot noir, but one of Virginia's newest….wineries— Ankida Ridge Vineyard—is opening my mind to the potential of still Pinot here in the Commonwealth."

Though I dropped the m-bomb—minerality—in my description of the 2010 Ankida Ridge Pinot, I no longer use that term since it is meaningless (see p. 107) Then there is the other m-bomb, masculine (see p. 316).

A lot has changed since 2012. The Wine Loft is no more. Virginia wine pioneer Dennis Horton died in June 2018. Both Oregon and Virginia wine industries continue to flourish. There are now over three hundred wineries in the Old Dominion. Though viognier was declared the official grape for the Commonwealth by the Virginia Wine Board, other whites, especially petit mansang, have been getting more attention. Virginia cabernet franc is still highly regarded but Bordeaux blends are burgeoning.

In January 2013, about one month after this piece appeared, I visited a handful of wineries in and around Charlottesville for the first and, thus far, only time. At Ankida Ridge, I tasted the 2010 and 2011 pinot noirs side by side. It was a revelation.

The younger wine had none of the jamminess of the 2010. I learned that this was because there wasn't enough fruit from the Vrooman's vineyard that year, so it was supplemented with pinot from Stinson Vineyards, a much warmer site, that was picked at a higher sugar level. The 2011 was from all-estate fruit.

After our visit to Ankida Ridge, we went to Veritas where we tasted a range of wines including a lovely ten-year-old chardonnay with proprietor Andrew Hodson. I would certainly go back.

We also visited Barboursville Vineyards. I was surprised at the poor showing of its 2009 Cabernet Franc Reserve during the tasting since I had enjoyed previous vintages. When I mentioned this to our host, he responded that was the May bottling and poured us the one from June. Quite a difference. I took one home.

Speaking of cabernet franc, Leah Jørgensen Jean had been making a name for herself specializing in that variety in Oregon, with fruit coming from the warmer southern part of the state and, in 2021, from the Red Mountain AVA in Washington State. She is credited with making the first white cabernet franc in the US. Really yummy.

Ankida Ridge's reputation continues to grow. In 2016, it was invited to participate in the International Pinot Noir Celebration in McMinnville, Oregon. Not having a pass, I stopped by to say hi. Christine slipped me a taste of the 2014 pinot, which was very nice.

I don't usually praise editors' choices for titles to my articles, but I have to hand it to Hilary Berg, then-editor of the Oregon Wine Press, *who came up with this gem in place of the insipid "Refusing to Pick Favorites" I offered.*

The Chosen None

Visitors to the tasting room where I pour quintessential pinot noir regularly ask me what my favorite wine is. Occasionally, my inner smartass causes me to blurt out, "Why, are you inviting me over for dinner?"

But my politer response is a homage to St. Anselm whose syllogistic ontological proof of the existence of God is just about the only thing that sticks with me decades after taking a medieval philosophy course as an undergraduate. "My favorite wine," I solemnly assert, "is the one in my glass at the time since it exists not only in concept but in reality."

Now well into my fifth decade of serious wine consumption, this declaration seems truer than ever. Having tasted thousands of wines, how could I honestly choose?

During the early 1970s, when wine was an occasional luxury, I kept detailed tasting notes of every bottle I sampled. (See p. 9) Nowadays, I rarely do so since wine is a part of my daily diet. Obviously, these bottles are not all created equal but, with few exceptions, they serve the purpose for which they are chosen very well.

Dinner parties with special guests merit the pick of my collection: grower champagne to start and a well-aged Oregon pinot noir or burgundy with the entrée, for example. Quotidian fare, such as a simple salad with canned tuna, is matched with a pinot gris or vermentino while ground mutton goes nicely with early-drinking Italian varieties, such as dolcetto. I must confess, however, that I frequently take advantage of the fact that the composition of my collection allows for more upscale pairings with a modest meal for just the two of us.

My aversion to picking favorites is likely rooted in my "satisficer" personality. In a 2002 paper entitled "Maximizing Versus Satisficing: Happiness Is a Matter of Choice" in the *Journal of Personality and Social Psychology*, Barry Schwartz and five colleagues build on ideas first published by Nobel Laureate Herbert A. Simon.

We learn: "To satisfice, people need only to be able to place goods on some scale in terms of the degree of satisfaction they will afford, and to have a threshold of acceptability. A satisficer simply encounters and evaluates goods until one is encountered that exceeds the acceptability threshold." As an "omnipour," I face no dilemmas.

While I easily resist the urge to evoke the F-word (*favorite,* that is), I learned that I have a "tell" that invariably exposes an exceptional vinous experience. Lately, many of these have been with my friend, the superb winemaker Jesús Guillén, who noticed that I emit an "oooh" when tasting something extraordinary, such as his 2011 Guillén Family Pinot Noir Reserve "Adrian" or the 1949 Volnay-Caillerets from Pierre Latour that we shared as part of my sixty-fifth birthday celebration in 2014.

Since each bottle (especially older ones) is unique, it would be fatuous to call it a favorite since it is unlikely to be exactly the same if one could taste it again. "Memorable" would seem to be a better adjective.

We are constantly encouraged to develop brand loyalty and many of us do. It is one thing to have a favorite pen or blue jeans but entirely another when it comes to consumables. Do we eat the same food every day or have the same rotation of dinners every week? I certainly don't. Should we be expected to drink the same selection of wine, forgoing the adventure of trying something new? I don't think so. When it comes to wine, brand disloyalty is much more appropriate.

What about favorite vintages? While I prefer those from cooler years, I certainly enjoy a well-made wine from any vintage. One of my mantras these days regarding recent back-to-back Oregon pinot noir vintages is "date the '12s,' marry the '11s.'" While I'm waiting for the complex, elegant, restrained 2011s to come around, why deny the obvious '12s'?

What about favorite grapes? Although pinot noir and riesling hold a special place for me, I don't pine for the former when I'm drinking a big-ass red with a hunk of grilled beef nor do I crave the latter when I'm indulging in a sauvignon blanc with a true cod.

Is one designating a favorite if one participates in ranking wines during a tasting? On its surface, that would seem to be the case since the bottle in first place has earned the right to be called the favorite. But this ranking is ephemeral since it is strongly tied to many factors, including the particular selection and order of the wines in the flight and the condition of the taster's nose and palate. A change in any of these can result in a reordering.

What is the point of picking a favorite anyway? Wine is a journey that should only end when you do. Selecting a wine above others can lock you into something that is very likely to change over time.

The minute a wine is barreled, it develops its own personality, hence the preponderance of Best Barrel, reserve, and winemaker's cuvées. Each bottle undergoes an individual metamorphosis, so it is virtually certain you can never taste the exact same wine twice. Better to savor what is in front of you rather than wish it were something else.

So, next time you are tempted to ask what my favorite wine is, ask, instead, in your best Samuel L. Jackson voice: "What's in your glass?"

I continue to be asked the question and show this piece as my answer. It does the trick.

Kaleidolfactic

Wine writers frequently attempt to communicate the impact of their favorite beverage on their senses through a stream of nouns, adjectives, and adverbs. The *McMillan Dictionary's* first definition of communication is "the process of giving information or of making emotions or ideas known to someone." For those willing to engage, wine is a rare, if not unique medium, for this process. But wine isn't only a medium of communication but also a message. At its best, it is a pure vinous statement of a particular place at a particular time, expressed via a particular winemaker.

Natalie MacLean, a marvelous mistress of metaphors and splendid sage of similes, declares, "Wine is the voice of the soil; without it, the land would be silent."

Unlike other types of art, such as painting and music, wine must be assimilated into several of one's senses to be appreciated fully. At its worst, a characterless wine is as engaging as a bore.

A bottle of wine is a time capsule with a liquid core. A mere whiff of the 2000 Domaine du Pegau Châteauneuf-du-Pape Cuvée Réservée transported me back to my early days of tasting, when I drank many remarkable older European wines, in a way that neither old photographs nor detailed tasting notes could. It was a pure example of an olfactory stimulation triggering vivid memories. And yet, I would be hard-pressed to record in detail the various components of the ever-changing bouquet.

I am now unable, or perhaps unwilling, to record my impressions as most others do with a list of very specific fruits, flowers, spices, minerals, and other flavors and scents. This runs contrary to my practice during the 1970s when I recorded detailed impressions of every wine I tasted (see p. 9). It also, ironically, flies in the face of my motto: "Reducing Vagueness since 1949." Nowadays, I feel that the experience of smelling and tasting wine and communicating its impact justifies keeping things somewhat vague.

Mathematicians are prone to fits of abstraction. As we contemplate conjectures or seek new theorems, our consciousness enters a different realm. So too can be the experience of savoring wine. While mathematicians have developed their own language that includes unambiguous definitions and sets of symbols to precisely convey ideas, such cannot be the case for wine tasters.

Instead, the attributes—the structure of a wine's message—must be described with imprecise words yielding, at best, a crude semblance of what is being communicated. This description offers what math folks call a "linear approximation" to the experience, meaning that they can only convey a rough idea of how a wine smells and tastes.

Translating the aroma or bouquet of a wine to words ranges from the obvious (a Rogue Valley 2009 Grenache really did smell like ginger) to the absurd ("On the nose, Albion strawberries picked by virgins at first light two days before the summer solstice complicated with elusive aromas of *cañela*; diaper of a three-month old, breastfed male child; Venus flytrap; and a half-smoked, pre-Castro Havana robusto"). But in general, I eschew orgasms of such descriptors. In my opinion, as is the case with music, to convey the vinous message, words fail.

Nevertheless, while sniffing a particularly lovely pinot noir—whose charms were revealed in a continuously evolving array of aromas—the specific associations with flora, fauna, fungi, forest floor, and spice box were less interesting to me than the pleasure of experiencing how things changed.

I then decided to make a virtue of vagueness and coin a neologism that captures this evolution: "kaleidolfactic." My definition of this adjective: exhibiting ever-changing aromas, as in: "Of all the red wine grapes, pinot noir is the most kaleidolfactic."

Linguistic purists will cringe. "Kaleidolfactic" can be viewed as a bilingual frankenword with roots in Greek—*kalos* which means "beautiful," and *eidos*, which means "form;"—and in Latin, *olfactorius*, from *olfacere*, "to smell." Yet the word feels good to say and fills a need to describe, in less precise terms, what one, with nose in glass, senses over time. It also does justice to the definition of communication.

I'm still the only one to use the term, but at least it's been included in some of my published wine descriptions. Maybe someday it will catch on.

In Defense of Describing Wines as Masculine, Feminine, and Sexy

Except for my own personal use, as a favor to a friend or colleague, or to satisfy a requirement for a gig, I eschew writing wine-tasting notes. Consequently, I dismissed Vicki Denig's rant against alleged sexist terms on wine-searcher.com on October 20, 2020 (https://www.winesearcher.com/m/2020/10/time-to-kill-gender-stereotypes-in-wine) as yet another misguided lunge by a hypersensitive.

But when it became the subject of an entire session entitled "Term Exploder" on the first day of the Symposium for Professional Wine Writers (WWS21, held via Zoom from May 10–12, 2021), my reverie was disrupted, and I was rudely awakened. The cancel culture has seeped into the world of wine writing. In response, I took to the chat to offer a different perspective. I present this rebuttal based on the position I put forth in that chat.

At the start of the session, the panelists were asked to "Explode this Tasting Note:"

"A wine of great breeding, the XXXX bursts from the glass with sweet smells of black currant, *pain grille*, and exotic spices. Masculine on the palate, with a sexy core of rich, dark fruit supported by a lingering acidity. Has the potential for medium-to long-term cellaring and would pair well with almost any stewed meat dish. A serious wine for the collector set and a fine example of the varietal."

Almost every adjective and noun pushed someone's buttons, with "masculine" and "sexy" singled out for extensive condemnation. Who knew the path from wines to lines could be so fraught?

This session elicited responses from two admittedly more notable wine writers. In her article, "The evolving language of wine" (https://www.jancisrobinson.com/articles/evolving-language-wine), Jancis Robinson writes: "I guiltily did a quick search of the 200,000-plus tasting notes published on JancisRobinson.com since 2000 and—sure enough—found 192 masculines, 147 feminines, and 37 sexys, although many of them were quotes from producers, or were preceded by the get-out 'stereotypically.'"

Without an ounce of guilt, I decided to scan through my four hundred fifty notes on wines I sampled between 1969 and 1979. (see p. 9) I found three that contained "feminine" and none with "masculine" or "sexy." (More on how I've been making up for this omission lately below.)

Here is part of my description of a 1962 Château Margaux that I tasted on October 2, 1977: "… Lovely medium deep elegant mature color. Flowery perfume—vegetable bouquet prominent at first—with air—nose becomes better balanced—flowery, fruit, herbal. Delicate flavor—flowers and fruit fade rapidly into a lovely long finish. Very feminine. Overpriced [at $27.50 less 10%, mind you], but interesting…"

My reaction to a 1967 Corton "Hospices de Beaune" consumed on January 12, 1976, concludes with "A very pretty, feminine burgundy." And then there is a 1970 Gevrey Chambertin sampled on November 7, 1975: "…Light, elegant, well-balanced taste—very feminine taste."

Now, so many years after they were written, these records of wines help me recall the experience of drinking some truly exceptional bottles. Until recently, I would engage in a parlor game with my dinner guests and ask them to read a description I had written decades earlier to see if I could recall which wine it corresponded to. Gender terms are among those useful in stimulating such memories.

W. Blake Gray blogged his reaction to WWS21 under the heading "Professional wine tasting notes are for the reader, not the writer" (https://blog.wblakegray.com/). A long-time hater of sessions on tasting notes, Gray offered a two-part rant focusing on the purpose of describing a wine in words.

While I appreciate his complex and nuanced arguments, I take issue with the following: "Nobody should call a wine 'masculine' or 'feminine' in 2021 because nobody knows what that means anymore; half the women in San Francisco can kick my ass, and the other half say, 'What do you mean, only half?'"

I certainly have no trouble knowing what masculine and feminine mean in the biological sense and have an unambiguous notion of what I mean when describing wines with these terms. Also, there are plenty of wine terms being used that have no universally recognized meaning.

For example, consider the pervasive "minerality" which carries with it the additional absurdity that rocks have taste or smell. Instead, what we are doing here is using the terms as metaphors, which can evoke memories of similar tasting experiences. They are certainly not intended to be offensive or to be in any way exclusionary.

The latter was the justification given by the panelists for retiring these terms—without any evidence, anecdotal, or statistical—that folks are traumatized by their use. Certainly, men enjoy wines described as feminine just as women enjoy wines described as masculine.

In an inane conflation, Denig advises: "Next time you're tempted to use a gender-focused tasting descriptor, think about how you would react if someone characterized a wine as 'white/Black,' 'gay,' or 'elderly' on the palate. If you'd find any of these terms offensive, then imagine how some of us men and women feel."

I'm sorry, I simply don't buy into this comparison and even find it offensive.

I remain unchastened. In fact, I have since increased my use of these terms and even found a way to acknowledge those who haven't made up their minds about which sex they are.

At one of the tasting rooms in which I pour, there is a wine that naturally lends itself to being described in gender terms. It is a lovely pour that starts masculine, i.e., rustic and funky, then gets in touch with its feminine self, exuding floral and perfumed aromas, before returning to show its more macho side. This single vineyard pinot noir is a shining example of a gender-fluid fluid! Far from offending visitors, my characterization is appreciated, revelatory, and even endorsed. No one has pushed back, and sales are good for this higher-priced bottle.

Denig made this offer to those who might be offended: "Next time a winemaker, tasting room employee, or sommelier uses a gender-focused descriptor, feel free to check them. Or send them my way." I look forward to her call.

"Sexy" also came under attack. One of the WWS21 panelists termed it "awkward." But once again, these PC word police have arrogated the responsibility to purge the language of descriptors that they deem inappropriate, without offering any evidence of the need to do so beyond their feelings or the feelings of those they seem to want to represent.

But since "sexy" is used to describe a particularly alluring or seductive bottle—without any reference to the various facets of the act like who, how many, what, what kind, where, how often, and with which parts—the word should remain in the lexicon of terms. One is free to ignore the term or use his or her imagination to personalize its meaning.

"Slutty" also came up and in the heat of battle, I agreed in the chat that this was an unacceptable term. I hereby withdraw my objection. I have, in fact, had wines that were overly generous and a little too eager to please.

Like Denig, the same panelist who had problems with "sexy," labeled "masculine" and "feminine" as "lazy cliches." He was joined by his fellow scolds. But like all imprecise descriptors, really the preponderance of those used for wine, they are merely suggestive and can elicit memories of similar wines.

If you want to attack a term for being lazy, look no further than the aforementioned "minerality," the pandemic use of which has led

Alex Maltman, a noted Welsh geologist and winegrower, to produce a stream of articles and a book (see p. 107) to set straight the record.

It is also a term for which there is no consensus definition. Everyone seems to acknowledge, and science provides solid evidence, that one's perception of wine is subjective. Compound that with different cultural references and experiences, and no one can expect anyone else's tasting note to precisely reflect his or her perception. Furthermore, tasting a glass of fine wine over a period of time is like dipping your feet into a stream. It is never the same moment to moment.

And what about wine scores? Despised by many but used, nonetheless. Even WWS21 keynoter Jancis Robinson expressed her disgust with them yet still assigns them. As an applied mathematician, I regard scores as a most egregious form of number abuse, ironically referenced with reverence by innumerates! Should I start a movement based on my bruised sensibilities to ban their use? Better to simply ignore them.

While free speech is a precious right, there is no inalienable right not to be offended, especially on behalf of unnamed others. As such, I am not particularly interested if you find my terminology lazy, inappropriate, noninclusive, or dated. It works for me and likely others who use it or resonate with it. If you can't stand the reference, take heart; many of us are boomers who are slowly leaving the wine scene. I hate tasting notes anyway. What these verbal prohibitionists are advocating is a one-size-fits-all version that will certainly make them so diluted that they become even more useless. Nevertheless, this free speech absolutist welcomes all voices in wine writing and believes that all should be heard…including mine.

Now go 'way and let me nap.

Don Kavanaugh, editor of wine-searcher.com, *to whom I offered the piece, replied, not unexpectedly: "I won't be using it, but thanks for an entertaining read!" It was also turned down by SevenFifty Daily. So I posted it on* medium.com. *Later, a shortened version was published in the* American Wine Society Wine Journal.

The response was generally positive, with one notable exception. From a Canadian sommelier and restaurateur: "In defence of toxic masculinity… Yes, definitely take me off your mailing list. Our belief are (sic) 100% not lined up."

Then there are these: "The best article you have done, almost EVER!! Not only did I find it well-written and amusing, it was spot on!! Cheers to you and I'll take a sexy, seductive wine any day!"

From another friend to my right: "What a truly wonderful article you wrote!! And no surprise they would not publish it. Though you and I disagree on many implementation details of a decent society, we both believe in our hearts in the First Amendment."

At the suggestion of wine writer and author Richard Leahy, who was also a fan of the piece, I submitted it to the Circle of Wine Writers for inclusion in The Circular, *the online publication accessible only by members. It was quietly ignored.*

In January 2023, Meg Maker, an accomplished wine writer and illustrator, led a panel discussion at Unified Symposium entitled "A New Lexicon for Wine." The panel included only women, no cranky old guys. She followed up with an article that got top billing on Wine Business Daily *(https://terroirreview. com/2023/02/02/we-need-to-talk-about-wine-talk/) and in May that year gave a seminar on the same subject to the Circle: https://www.youtube.com/watch?v=00dtqAOi3bY.*

I was particularly amused by Meg's characterization of the gender terms as "othering," whatever that means. In any case, in the spirit of collegiality, I had sent my article to Meg in advance of her presentation—which she politely acknowledged receipt of— behaved myself during it, and even followed up by e-introducing her to Patrick Reuter, winemaker and co-owner of Dominio IV, who does "shape tasting" (https://www.oregonwinepress.com/the-shape-of-wine). She seemed to resonate with the idea.

Gendering wine has been a topic in academic literature for decades. In a graphic representation of terms relating to a wine's *balance, the revolutionary French oenologist, Émile Peynaud, shows a masculine-feminine axis (*The Taste of Wine, *1987, The Wine Appreciation Guild, p. 171).*

In a paper in the Journal of Wine Economics, *Philippe Masset, Lohyd Terrier, and Florine Livat ask: "Can a wine be feminine? Gendered wine descriptors and quality, price, and aging potential" (https://doi.org/10.1017/jwe.2023.30).*

They conclude that wines described using feminine terms are perceived as not as capable of aging as those with masculine descriptions. They found no difference in wine scores or pricing, nor did they find that the use of feminine wine descriptions increased over time. This should bring some comfort to those sensitive souls.

Wine-searcher.com declined to publish this, so I posted it on medium.com.

Requiem for My First Wine Friend

John, whose fault it is (https://www.wine-searcher.com/m/2020/06/the-cost-of-drinking-wine-history), died of a massive heart attack while walking on November 19, 2022, twelve days short of his seventy-third birthday. I met Toshu John Neatrour in August 1968 when he allowed me to stay at his off-campus apartment until the dorms opened late the following month after I came to the area for an Astronomical League Convention in Chicago. We were both astronomy majors at Northwestern University in Evanston and lovers of classical music, so there were already common pursuits.

But it was John who was singlehandedly responsible for igniting what became my lifelong obsession with fine wine. Yet, over time, this erstwhile orchidist's, avid telescopist's, and onetime semiprofessional cellist's interest in the spirituous world, while never diminished or never blossomed to the same extent as mine but, instead, was eclipsed and came to play second fiddle to one more spiritual. (John was also my most formidable "Pun Pong" opponent.)

John was attracted to Sōtō Zen Buddhism in the 1970s while we were both graduate students, even bringing me along to practice zazen at a temple in Chicago. After a few years, I stopped meditating, but later he was ordained and changed his first name from Marvin to Toshu. The Dragon Chant Zen Center that he shepherded became the focus of his life until the end.

Over the first decades of our friendship, our educational and career paths diverged leading to geographic separation. We visited intermittently, sometimes with gaps of a few years. But then, in 2002, we ended up in the same company on the West Coast. Later, he joined me at a professional services company on the opposite coast. When we left full-time employment, we once again lived in separate but this time contiguous states. We stayed in touch and even collaborated on two papers that were published in peer-reviewed journals.

While each of these adventures could fill pages, it is the oenophilia that I contracted from him, along with the attendant gluttony, that is the source of my most vivid, albeit fading, and pleasant reminiscences. So, the most fitting eulogy for this lapsed meditator to give to my first and dearest wine friend is a recollection of some of the bottles we shared over the years.

When I first met John, he had two classified growth clarets, 1961 Leoville Las Cases and 1961 Calon-Ségur. The first he recalled us drinking with his mother's pasta sauce that he prepared, but the circumstances of the uncorking of the second remains a mystery. I didn't start keeping notes about the wines I tasted until August 1969, so it is likely that if I had the Calon- Ségur, it was before then.

My notes during the first decade didn't contain the names of those who shared the wines with me, but there are some hints from the circumstances of their consumption when included. John was a founding member of the Duncan Hines Memorial Bon Vivant Fellowship, International, or DHMBVF, the gourmet society I cofounded in 1969. The tastings and annual banquets featured old bordeaux and burgundies, the occasional California wine, and German bottlings. Two such events that took place in November 1975 have been previously recounted (https://www.wine-searcher.com/m/2020/06/the-cost-of-drinking-wine-history).

A tasting of Medoc reds was held in my apartment on December 23, 1972, with John; his first wife, Alice Berkson; Jim Kettner and his wife, Janet; and fellow oenophile and astronomy major, Spencer Young, who kept a sizable cellar at his parents' house in Lagrange, Illinois, from which we enjoyed many a bottle.

The flight comprised 1961 Leoville Las Cases, 1966 Château Latour, 1953 Cos D'Estournel, and 1934 Château Margaux. The 1961 Château D'Yquem was dessert.

John assisted me in the five-day preparation of a dinner celebrating my master's degree on June 17, 1973, at which a 1970 Château Carbonnieux Blanc, a 1947 Chambertin from Pierre Ponnelle, and a 1969 Hattenheimer Heiligenberg Feine Beerenauslese were poured.

After John finished his bachelor's degree at Northwestern, he switched to physics for graduate school at the University of Illinois, Champaign-Urbana campus. His master's dinner and early twenty-fifth birthday celebration on November 2, 1974, included the 1949 Chambertin Clos de Beze from Bouchot-Ludot, a 1947 Clos de Vougeot from Pierre Ponnelle, the 1959 Leoville-Las Cases, J. J. Prüm's 1949 Wehlener Sonnenuhr Feinste Auslese, and a 1971 Eltviller Sonnenberg Riesling Beerenauslese. Moderation had yet to enter our vocabulary, let alone our habits.

On May 22, 1977, we celebrated finishing my PhD with a tasting in my small apartment in graduate housing with some faculty members and their wives. The lineup included the 1959 Château Beychevelle, 1962 Château La Mission Haut Brion, 1959 Château Mouton-Rothschild, and a 1959 Château Talbot, along with a 1971 Steinberger Spätlese and the fantastic 1959 Steinberger Trockenbeerenauslese.

I moved my tiny collection of bottles to Nashville, Tennessee, all of which were consumed during the year-and-a-half I was there attempting a career in academics. I'm sure that John was involved in consuming at least one when he visited, but I have no record. Not making enough to support my wine habit—not that there was much to buy in town—and the family that I had, I moved to California in 1979 to become a rocket scientist. John remained in the Midwest, eventually settling for an ABD (all but dissertation) and working as a software engineer for several organizations.

In addition to our careers that put us in different parts of the country, our marriages, children, "unmarriages," and new marriages kept us occupied. Visits were rare while the lack of modern communication methods like email and mobile phones necessitated letter writing, a more time-consuming and, hence, rarely utilized method to keep in touch.

In 1984, I was once again in Chicago for a non-work-related meeting when I got to see John. I think it was an Alsatian wine (or two?) that accompanied the over-the-top oyster extravaganza lunch in Oak Park. To this day, it is the single most expensive midday meal I've had.

Tiring of California and enticed by a sweet offer in Massachusetts, I moved with my family across the country in 1988. John remained in the Chicago area until 1999. That same year, I bounced back to the West Coast—this time, Redmond, Washington—to try my hand in the software business.

In the first years of the new millennium, John pursued Zen matters full-time in Japan. In 2002, he was in Milwaukee. That was also the first year for which I have our email exchanges and the year he came to work for the same software company.

With this renewed proximity, though short-lived, and ample funds, we feasted and drank mightily and well. The now-defunct Fine Wine and Cigars in Redmond Town Center was the site of many a Friday afternoon tasting and the source of truly fine bottles. Woodinville, a vineyard-less nascent center of winemaking, was a short distance and was starting to grow. Of course, my wine stash began to veer heavily toward Washington State bottles.

During the less than two years we were together in Redmond, there were many dinners at my home for which, unfortunately, I have no records. Both Madeira lovers, John and I did enjoy a dinner at Matt's Rotisserie & Oyster Lounge, at that time, which featured a few from the nineteenth century.

After a change of fortunes in 2004, I ended up in Northern Virginia. John was to follow the next year, accepting a position in the professional services company I was already working for. Records of dinners and tastings during that period exist.

In May 2008, I hosted a dinner for John and another colleague that featured a nonvintage (NV) R. Dumont & Fil Brut and 2000 Château Duhart-Milon. At another, less than two months later, an NV R. Dumont & Fil Brut Rosé, the 2000 Te Mata Estate Coleraine Cabernet/Merlot, 2000 Hartwell Vineyards Mistique Cabernet Sauvignon, and the 2005 Wehlener Sonnenuhr Riesling Auslese from Cardinal Cusanus Stiflswein were served.

Two weeks later, I poured NV R. Dumont & Fil Blanc de Blancs Brut, the 2001 and 2002 Betz La Serene Syrah, and a Marc de Chateauneuf de Pape.

John and I attended a NASA-sponsored conference in Portland, Oregon, in 2008, and celebrated forty years of friendship with a dinner at the now-shuttered Heathman Restaurant. It included a Sancerre and a 1998 Ken Wright Cellars Pinot Noir. We were impressed with the decade-old pinot.

Occasionally, we would get live Maine lobsters off a truck that came down to Virginia. A dinner featuring the crustaceans in October 2008 was accompanied by an NV Rondel Pura Raza Cava, the 2002 Allenberg de Bergheim Riesling, the 2006 Bergström Sigrid Reserve Chardonnay, and the 1997 Château Suduiraut.

Virginia wines began to entice and found a place at some dinners. In February 2010, NV Thibaut-Janisson Brut Rosé led one off that also included a 1999 Domaine du Pegau Chateauneuf du Pape.

On March 18, 2010, I prepared a dinner to celebrate the publication of "The Judgment of Paris According to Borda" (https://doi. org/10.1080/09571260903451029), in which I demonstrated that the 1970 Château Haut-Brion actually edged out the 1973 Stag's Leap Wine Cellars Cabernet Sauvignon for first place. I had managed to procure one of the last bottles of the former and served it.

John's reaction: "It was wonderful. Under a constant theme of cedar, various flavors evolved. A general red fruit became more distinctly cherry and then cherry-raspberry. A mineral-herbal complex provided the most distinct entertainment, showing itself as earth and grass, then with added mint notes, shifting toward damp straw, horse sweat, and rain on rocks."

The celebration began with the NV Thibaut-Janisson Brut, included the 2001 Clos de Betz for contrast, and ended with the 1994 Dow's Vintage Port.

The last dinner John attended at my home in Virginia, for which I have a menu, was on February 9, 2011. The NV Agrapart & Fils Les 7 Crus Brut Blanc de Blancs, a 2006 Remy Wine Lagrein, and the 1985 Sandeman Vintage Port accompanied the meal.

During the summer of that year, I was converted to a consultant and no longer needed to live in Virginia. We then became Oregon residents while John stayed in the Commonwealth until 2019 when he moved to Idaho.

He visited us in Oregon twice. More notable is the second time in 2014 when he came to celebrate my sixty-fifth birthday. In addition to John, Jesús Guillén, the winemaker at White Rose Estate and his own label, Guillén Family Wines, and his wife, Yuliana, joined us. (After the tragic early death of her husband (see p. 230), Yuliana assumed control of the brand and oversees the production of some lovely pinot noirs.)

The dinner started with the 2007 Riesling Kaeffererkopf from Meyer-Fonné and was followed by a 1949 Volnay-Caillerets from Pierre Latour, which Jesús said was in the top fifteen of about three thousand wines he tasted. In case that bottle was bad, we had as backup, but consumed anyway, a 1999 Archery Summit Arcus Estate Pinot Noir and a 2006 Clos de Vougeot from Domaine François Lamarche. The former was one of Jesús's epiphany wines. Dessert was accompanied by an NV Phelps Creek Vin Doré Gewürztraminer (sic).

The last time I saw John was in September 2018, during one of our frequent trips to Virginia to visit our daughter and her family. We marked fifty years of friendship at L'Auberge Chez Francois in Great Falls with a 2014 Château Carbonnieux Blanc and the 2008 Château Malartic-Lagravière.

Despite moving closer to each other, we never were able to get together again. I am profoundly saddened that there wasn't at least one more chance to share a fine bottle or two with John. I had hoped to celebrate his seventieth birthday with him in 2019 at my home in Oregon, but circumstances and distance prevented it.

I was planning to serve three pinot noirs from the 2008 vintage, exactly forty years after we met, a White Rose Estate White Rose Vineyard Whole Cluster (disclosure: I work part-time in the White Rose tasting room), the Archery Summit Archery Summit Estate, and the Château de la Tour Clos Vougeot (also 100-percent whole-cluster). They will remain on my wine bucket list unconsumed (see p. 330) until there is a worthy occasion and appreciative company.

I was also saving an NV Laherte Frere Les 7, a rare grower champagne made from the seven permitted varieties. Then the global insanity intervened and finally his death, tragically precluding forever the visit once contemplated for the very same month he died.

So now, there are only receding imperfect recollections of our times together and apart over more than half a century through school, marriages, children, diverging and intersecting careers and passions, collaborations, and occasional disagreements. But the sweetest memories of all flow from recalling the wines we shared.

Thank you, John, and rest in peace.

My Wine Bucket List

I'm not one to practice being old. Whenever asked where I grew up, I reply that I never did and don't plan to. There is simply no benefit to doing so. I insist, as did my friend John, that I'm nineteen in my head (though I really should say twenty-one lest I run into trouble with the "thought police" for underage drinking).

But the reality is that my body has a mind of its own and insists on calling the shots. It's been lobbying, thus far unsuccessfully, to be alcohol-free. So now, into my eighth decade, I have been thinking about what I'd really like to do before I can no longer do anything.

I have visited six continents and want to complete the set before the seventh one completely melts. I still have three US states to set foot in and many more countries to visit. Of course, there are books to read or reread, movies to see or see again, and concerts and operas to attend. And then there are wines.

Something Old

For a long time, I envisioned expiring while sitting at my desk doing math with a glass of ancient vintage port, then having my ashes scattered in a vineyard. These days it is more likely that I'll become a hood ornament on a vehicle operated by someone driving while self-absorbed or texting as I cross the intersection of 2nd Street and

99W in my hometown before my pump has a chance to peter out on its own.

Also, my wife is a firm believer that the couple who stays together should decay together. So when we irreversibly plotz, there are side-by-side plots in a small quiet Jewish cemetery in Albany, Oregon, that will become our home. But the desire to have a few more old ports, as well as other venerable vinos from the past couple of centuries, persists.

I have previously written about the four notebooks (see p. 9) that contain my impressions of wines I tasted from 1969 to 1979 (https://www.wine-searcher.com/m/2020/06/the-cost-of-drinking-wine-history). I drank more fortified and dessert wine back then than I do now.

Of the ones that I would like to revisit, which might still be around and, more important, still enjoyable, are the 1960 Vintage Port from Manoel D. Pocas Jr. and the 1960 Warre's Vintage Port. I am curious to see how the 1963 Quinta do Noval Port developed after noting, in 1977, that it was "Very much a fetal wine." And then there is the 1882 Blandy's Bual Madeira, which should evidence that madeira lives forever.

Other gems from my days of yore are a 1952 Châteauneuf-du-Pape from Pere Anselme, which I described as "A classic, probably not to be duplicated;" the 1961 Château D'Yquem; the 1959 Steinberger Trockenbeerenauslese; the 1959 Château Mouton-Rothschild; and a 1949 Wehlener Sonnenuhr Feiste Auslese from J. J. Prüm.

The last one was from my birth year and proved that not only was 1949 a great vintage for people but also for European wines. I have been able to celebrate a few of my birthdays for years ending in zero or five with bottles from that vintage.

The most recent, a 1949 Marchesi di Barolo Barbaresco, was well past its "best by" date in 2019 and well into its sipping vinegar phase. In contrast, the 1949 Volnay-Caillerets from Pierre Latour proved the hit of my sixty-fifth birthday celebration. It showed plenty of life and would easily have made it five more years.

For the next milestone birthday, I would like to have something of at least that quality from Bordeaux, perhaps Château Margaux, or another from Burgundy, the Domaine de la Romanée-Conti Romanée-Conti, and a DRC La Tâche would do. I also wouldn't refuse a 1949 Barolo Riserva Monfortino from Giacomo Conterno or a 1949 Blandy's Vintage Bual.

I could go on, but it would be an exercise in futility since all of these have been priced well out of my range.

Something New

So, let's be less aspirational and more practical and look at some affordable wines that aren't as old and that I have wanted to try. Over the past decade, I've become a fan of grower champagnes, which I exclusively get directly from Caveau Selections. I have a nice varied and varying selection that includes bottlings from Laherte Frères and Marc Chauvet.

Within the past few years, however, Scott Wright, former proprietor of Caveau, had been importing examples from new producers, which for various reasons, including limited space and limited availability, I have not purchased. E-dangled temptations include bubbles from Champagne Étienne Sandrin, Champagne Heucq, and especially Champagne La Parcelle.

Caveau also directly imports burgundies. Again, for reasons of space and availability but also in some instances, price, I have not acquired chablis from Fred & Céline Gueguen, Chapelle-Chambertin, or Corton-Perrières, both made by Wright under the Caveau label, or Grand Crus from Domaine Heresztyn-Mazzini and Domaine A-F Gros.

White Rhône and Italian wines have been rare pleasures. I would like to taste more bottlings from St. Joseph and Châteauneuf-du-Pape and all of the indigenous white Italian varieties from around the boot.

I occasionally sample the products of some of the newer American wine regions like Texas and Michigan and would like to delve deeper. I even hear there is some good stuff in Arizona.

Something Near

I needn't leave home to finally get around to drinking some wines that I've been wanting to. My collection has always reflected the region I either lived in or close to. So now, it includes a disproportionately large number of Willamette Valley pinot noirs as well as some chardonnay and other varieties from around Oregon.

Oregon pinots, especially those from cooler vintages, can age for decades. I plan to open a 2002 Delara Pinot Noir from Maysara, the oldest pinot I own, later this year. There are also quite a few others that I have cellared for over ten years.

For example, I have been saving the 2008 White Rose Estate White Rose Vineyard Whole Cluster Pinot Noir (disclosure: I work part-time in the White Rose tasting room) and the 2008 Archery Summit Estate Pinot Noir to taste alongside a 2008 Chateau de la Tour Clos Vougeot (also whole-cluster).

At the beginning of 2021, I began drinking 2011 pinots from the last cool year in the Willamette Valley, some of which could use more time. (See "The Value of Elegance," p. 293 for some standouts.) Fortunately, I have more.

There are still a few Oregon pinots from the 1980s and 1990s out there that I would like to try that aren't part of my cellar. I know that The Eyrie Vineyards, one of the two oldest wineries in the Northern Willamette Valley, keeps a library. There are other producers who do as well.

Also in my collection are bottles that I have set aside for grandchildren, vintages 2013, 2015, 2016, 2017, 2019, 2021, and soon 2024. These include magnums of 2013 Adrián Pinot Noir Reserve from Guillén Family Wines and 2015 RR Wines Ridgecrest Vineyards Pinot Noir. There are also magnums of 2017 and 2021 Laurène from Domaine Drouhin Oregon.

Through 2017, I have laid down best barrel bottlings of pinots and chardonnays from my former employer, Chehalem Winery, and a remarkable 2017 White Rose Estate White Rose Vineyard Pinot Noir. The hope is that I am around to enjoy these with the grandchildren either when their parents allow them a taste or when they no longer need to ask for permission.

Lest you think that I only collect pinot noir, I look forward to pulling the corks of a 2010 Armada Syrah from Cayuse Vineyards and the 2005 Paramour, a tempranillo in the style of a Gran Reserva, from Abacela, when the moment is right. I've just cellared the 2020 RR Estate Reserve Riesling and 2021 Ridgecrest Old Vine Pinot Gris, both made by Ribbon Ridge Winery, which I hope to enjoy when they are at least ten years old.

Something Far

Since 2012, I've attended most of the annual meetings of the American Association of Wine Economists (AAWE), which have been held— ironically, given the name—more often outside of the United States in fabulous places like Stellenbosch, South Africa; Mendoza, Argentina; and Padua, Italy. There is still more travel to wine regions I need to do, with or without the AAWE justification. I have yet to visit wineries in Eastern Europe, the UK, Israel, and China, among many others.

I would like to return to Australia and tour some regions there, not having had the opportunity to do so during a business trip in 1994. There are so many wineries I would like to revisit or visit for the first time in New Zealand, like Ata Rangi for its excellent pinot noir, and in South Africa, including Kanonkop for extraordinary pinotage (yes, there is such a thing), and Klein Constantia for a variety of expressions of sauvignon blanc.

If there is to be a second trip to South America, I would like to make a stop in Bolivia for the first time and taste more of their ethereal high-elevation whites. It should also include my first visit to Brazil for an immersion in bubbles. My second visit to Chile would focus on old vine wines and whatever Alfonso Soto Gonzales, my guide the first time we visited in 2015, recommends. My second visit to Uruguay must include a stop at Bodega Garzón.

In my coverage of the 2014 edition of the Passport to Pinot (see "IPNC for the Rest of Us," p. 155), what I have called the CliffNotes version of the International Pinot Noir Celebration, I was particularly taken by an Argentine offering:

"The 2012 Barda from Bodega Chacra in Northern Patagonia was something of a revelation, with funky aromas competing with floral notes and an elegant palate with lots of acidity and good tannins. Winery proprietor Piero Incisa, who showed interest in my connection to the American Association of Wine Economists, promised to stay in touch and could become my latest friend in the industry."

So far, it hasn't happened, but hope springs eternal that my next trip to Argentina will include a visit to the bodega and another encounter with Incisa.

Reality and an Offer

Prices of many of these wines have become prohibitive and are well beyond what I can afford. Andrew Jefford suggests: "Perhaps there might be a future for something called wine philanthropy." (https://worldoffinewine.com/2021/09/10/la-tache-hughess-hen-and-vermeers-virginal-wines-transactional-flaw/) If this becomes so, count me in on the receiving end. To anyone out there who will share one of these bottles with me, I offer a cameo in the video that I'm supposed to see before my mind's theater darkens for good.

Admittedly, this list is far longer than I'm likely to be around. So realistically, I will only be able to check off a few. Fortunately, I am a satisficer, not a maximizer, so whichever of these wines I'll be able to taste will be good enough. Truth be told, I would be delighted to try anything from anywhere I haven't before.

Since this piece first appeared, my friend, John Neatrour, died in November 2022 of a massive heart attack (see "Requiem for My First Wine Friend," p. 323). I consumed the 2002 Delara Pinot Noir from Maysara and the 2005 Paramour from Abacela with one of my sons in 2023. Both were memorable.

Also, in 2023, we returned to South Africa for the AAWE meeting, stopping on the way in London and visiting wineries in both places. (See "Back to Stellenbosch in 2023," p. 66)

In addition, I added Antarctica to my list of continents I set foot on. (There was still plenty of ice.) Now I have been to all seven. Though we stopped in Argentina on the way down, we didn't have the opportunity to visit any wineries, so a tasting at Bodega Chacra remains on the list.

Glossaries that I've seen in other wine books are routine, incomplete, and boring. As I began to populate this one, I realized that I might be able to avoid the old ennui. When I started, I knew that some of the definitions that first came to mind are certain to upset a few folks.

"Good," I thought, "it's my book, and I'll offend if I want to."

Then, I remembered Ambrose Bierce's Devil's Dictionary. *He didn't give a damn either about whom he insulted. Matt Kramer saw a similar opportunity when he published* The Devil's Wine Dictionary *in the November 30, 1995 issue of* Wine Spectator. *Here are definitions of terms that have gained prominence or notoriety over the last three decades, none of which appear in Kramer's list.*

The Devil's Wine Glossary

Feminine: a wine that exudes floral and perfumed aromas and sits elegantly on the palate. A term that rubs the professional offense-takers the wrong way, now gaining prominence in wine-writing circles, who label this and other terms (see **Masculine** and **Sexy**) "othering," whatever the hell that means. Best regarded metaphorically than literally.

Kaleidolfactic: a term I coined because I got tired of writers describing a wine's bouquet and/or palate as kaleidoscopic. The latter term refers to sight, not smell nor taste, and their derivative flavor. So far, no one has used it except me.

Masculine: a big, bold wine with giant balls and hair on its chest, also rustic and funky, that nevertheless can be enjoyed by any gender. Despite the temptation to take it literally, it works best as a metaphor. One of several terms (see **Feminine** and **Sexy**) that has become a target for wine-writer-scolds who closely affiliate with the cancel-culturistas and generally lack a sense of perspective, let alone a sense of humor.

Minerality: a bullshit term used pervasively, even mandatorily, that has no agreed-upon definition and certainly no basis in geology, yet makes some feel like insiders.

Numerical Score: the wine world's most egregious example of number abuse and, in some sense, the most ironic since it is craved by so many otherwise innumerates. An absurd bit of reductionism that, when above 90, nevertheless sells wine, thus converting its worthlessness into worth for savvy sellers

Sexy: yet another hot-button term (see **Feminine** and **Masculine**) that triggers anti-free speech rants from the humorless who would denude our language of all but neutered terms. It refers to a particularly alluring or seductive bottle, without any reference to the various facets of the act such as who, how many, what, what kind, where, how often, and with which parts. Anyone who thinks it does has a very dirty mind.

Slutty: still another hot-button term and one that even I once took offense at. But now, I don't. It aptly describes wines that are overly generous and a little too eager to please.

Tasting Note: an unfortunate necessity for a wine writer requiring him or her to perform an unending series of mental calisthenics and other gyrations to avoid repetition, or worse, the simple yet incontrovertible, "It tastes like wine!"

Wine Competition: a pay-to-play event in which producers gamble good money in the guise of entrance fees for a chance to win matte-finished medallions attached to colored ribbons and the right to put stickers on so-recognized bottles. Selections are made by **Wine Judges**.

Wine Economist: a practitioner of the dismal science who seeks to brighten his or her work by focusing on wine and other intoxicants. In reality, a flexible and fairly indiscriminate term that can be applied to anyone who affiliates with the American Association of Wine Economists.

Wine Judge: an individual who believes he or she has both the stamina and superior equipment to assess dozens of wines, yet more often than not cannot recognize the same wine served in more than one glass in the same flight. This, despite undergoing training, in many cases.

Wine Mathematician: a mythical being. Practitioners of the queen of sciences with an interest in wine are welcome to rebrand themselves as **Wine Economists** by joining the right group.

Chronology

All original texts have been revised and/or annotated by the author for publication in this book.

2024

The American Association of Wine Economists and Me (January) (page 29) - Not previously published.

Back to Stellenbosch 2023 (January) (page 66) - Not previously published.

2023

The Notebooks (page 9) (November) - Not previously published.

Rachising Up Pinot Noir (page 213) - *Oregon Wine Press* (November) pp. 36-41.

Linfield Raises the Bar (page 204) - *Oregon Wine Press* (November) as Acorn to Oak: Linfield Raises the Bar p. 12.

Willamette Valley's Way with Chardonnay (page 182) - GuildSomm.com (October 20) https://www.guildsomm.com/public_content/features/articles/b/neal-hulkower/posts/willamette-valley-chardonnay.

A Quick Look at Slow Wine (page 277) - *Oregon Wine Press* (September) p. 10.

Book Review: Drinking with the Valkyries (page 141) - *Journal of Wine Economics* 18 (March) doi:10.1017/jwe.2022.52.

Heavenly Juice (page 208)—*Oregon Wine Press* (March) p. 18.

Requiem for My First Wine Friend (page 323) - https://medium.com/@nhulkower/requiem-for-my-first-wine-friend-89b1c979031d (February).

2022

Georgia for the Wine (page 61) - *Oregon Wine Press* (October) p. 14 as Exploring Georgia.

A Post-IPNC Chat with the Master of Ceremonies, Philippe André (page 235) - *Oregon Wine Press* (September) as Philippe André p. 12.

Book Review: The Science of Wine, 3rd Edition (page 136) - *Journal of Wine Economics* 17 (August) pp. 264-267. doi: 10.1017/jwe.2022.37.

My Wine Bucket List (page 330) - wine-searcher.com (May 23) as Ticking Off the Wine Bucket List https://www.wine-searcher.com/m/2022/05/ticking-off-the-wine-bucket-list.

The Value of Elegance (page 293) - wine-searcher.com (March 28) as In Search of Elegant Wine https://www.wine-searcher.com/m/2022/03/in-search-of-elegant-wine.

2021

A Remembrance of German Wines Past (page 20) - *Trink Magazine* (September) https://trinkmag.com/articles/a-remembrance-of-german-wines-past.

Book Review: In Vino Veritas (page 131) - *Journal of Wine Economics* 16 (August) pp. 350-353. doi: 10.1017/jwe.2021.30.

Book Review: The Story of Wine (page 125) - *Journal of Wine Economics* 16 (August) pp. 347-350. doi: 10.1017/jwe.2021.29.

The Vineyard Stewards' Stewards (page 264) - *The Grapevine* (July-August) pp. 46-53.

In Defense of Describing Wines as Masculine, Feminine, and Sexy (page 316) - https://nhulkower.medium.com/in-defense-of-describing-wines-as-masculine-feminine-and-sexy-ebb35f6551a3 (July) and shortened version in the *American Wine Society Wine Journal* (Fall), pp. 31-32.

Graduation Day (page 260) - *Oregon Wine Press* (April) p. 9.

2020

Book Review: Pinot Girl (page 119) - *Journal of Wine Economics* 15 (November) pp. 436-439. doi:10.1017/jwe.2020.26.

AHIVOY Presses On (page 257) - *Oregon Wine Press* (November) p. 16.

Playing the Field for Slow Wine (page 273) - *Oregon Wine Press* (August) p. 26 as Zooming into Slow Wine.

Book Review: Wine Girl (page 113) - *Journal of Wine Economics* 15 (February) pp. 122-126. doi:10.1017/jwe.2020.10 also in the *American Wine Society Wine Journal* (Summer 2023) pp. 14-17.

There They Go! (page 252)- *Oregon Wine Press* (February) p. 16.

2019

Battle Creek Flows into the Pearl (page 202) - *Oregon Wine Press* (December) p. 16 as Battle Creek Rising.

Book Review: Vineyards, Rocks, & Soils (page 107) - *Journal of Wine Economics* 14 (May) pp. 217-220. doi:10.1017/jwe.2019.19.

Time's Way with Oregon Chardonnay (page 178) - *Oregon Wine Press* (April) p. 10 as Coming of Age.

2018

Jesús Guillén Olvera (1980–2018) (page 230) - *Oregon Wine Press* (December) p. 10.

Book Review: I Taste Red (page 100) - *Journal of Wine Research* 29 (August) pp. 225-228. doi: 10.1080/09571264.2018.1506323.

Breaking Down Walls (page 243) - *Oregon Wine Press* (August) p. 35-41.

Wine Economists Convene "Far above Cayuga's Waters" (page 49) (June) - Not previously published.

2017

In the Garden of Wine Economists (page 45) - *Oregon Wine Press* (August) p. 8.

VIDON to Release Space Exploration Series (page 226) - *Oregon Wine Press* (February) p. 8 as Space-Aged Offering.

The North, South, East, and West of Oregon Chardonnay (March) (page 174) - Not previously published.

2016

Grape Explications (page 285) - *Oregon Wine Press* (December) p. 26.

Book Review: Riesling Rediscovered (page 95) - *Journal of Wine Research* 28 (October) pp. 68-70. doi:10.1080/09571264.2016.1238351.

Remy Wines Takes to the Street for a Fair Celebration of its 10th Anniversary (page 223) - *Oregon Wine Press* (August) p. 8 as Remy Wines Celebrates 10.

The Day of Wine and Flowers (page 198) - *Oregon Wine Press* (July) p. 12 as Wine Country Canni-bus.

Grape Expectorations (page 289) - *Oregon Wine Press* (March) p. 22.

Fêting Chardonnay the Oregon Way (March) (page 169) - not previously published.

2015

Book Reviews: Winemakers of the Willamette Valley and Oregon Wine Pioneers (page 90) - *Journal of Wine Economics* 10 (December) pp. 382-384. doi:10.1017/jwe.2015.37.

A Random Walk through IPNC (page 159) - *Oregon Wine Press* (September) p. 24 as IPNC is for Connoisseur.

Don't Cry for Us: Wine Economists Meet in Mendoza (page 40) - *Oregon Wine Press* (July) p. 26 as AAWE in Argentina.

O Chardonnay (July) (page 165) - not previously published.

The Chosen None (page 310) - *Oregon Wine Press* (June) p. 26.

2014

Book Review: Best White Wine on Earth (page 85) - *Journal of Wine Economics* 9 (December) pp. 358-361. doi:10.1017/jwe.2014.35.

Kaleidolfactic (page 313) - *Oregon Wine Press* (October) p. 26.

IPNC for the Rest of Us (page 155) - *Oregon Wine Press* (September) p. 10.

Wine Economists Powwow in Walla Walla (page 36) - *Oregon Wine Press* (August) p. 12 as Following the Paper Trail.

2013

Passport to Pinot: Something to Walkabout (page 151) - *Wine Press Northwest* (Fall) p. 44.

A Stellar Bash in Stunning Stellenbosch (page 31) - *Oregon Wine Press* (August) p. 10 as Stellar in Stellenbosch.

Consensus Ranking of Oregon Pinot Noir Vintages: Clash of Sensibilities (page 54) - *Oregon Wine Press* (June) p. 23 as Clash of Sensibilities.

Book Review: how to love wine (page 79) - *Journal of Wine Research* 24 (February), pp. 160-162. doi:10.1080/09571264.2013.768215.

2012

Vindicating Thomas: Virginia Vanquishes Vinifera (page 298) - *Oregon Wine Press* (December) p. 28 as Vindicating Thomas.

Book Review: Voodoo Vintners (page 73) - *Journal of Wine Research* 23 (July), pp. 255-257. doi:10.1080/09571264.2012.701612.

2011

Borda is Better (page 16) - *Oregon Wine Press* (October) p. 35.

1974

The *Vintage* Tasting Notes (page 13) - *Vintage* (January) p. 46.

1973

The *Vintage* Tasting Notes (page 10) - *Vintage* (September) p. 53.

Acknowledgments

I am especially grateful to Donald G. Saari, my dissertation advisor and friend who, after becoming a fan of my wine writing, urged me to consider creating a book. After a few years of prodding (as in, "You're not getting any younger") and the accumulation of what I considered a critical mass of suitable material, I finally did it.

My wife, Clara, is my first and best editor and critic who, in her capacity as family CFO, had to approve the funds for this project. I am forever thankful for her support and encouragement.

Thanks to those who gave support, encouragement, and advice, especially the Association of Independent Authors, during my initial plunge into self-publishing.

The team at Luminare Press did a brilliant job of preparing the manuscript for production, from copyediting, cover design, formatting, and proofreading through establishing the interfaces with Kindle Direct Publishing and IngramSpark.

Special thanks to Courtney Cunningham of Glint Creative for turning my crude scribble into the lovely cover art.

I am thankful to the readers whose appreciation over the years made my efforts worthwhile beyond the sense of accomplishment and satisfaction it gave me.

Finally, a big thank-you to all of my subjects who gave me something interesting to write about.

Index

351

M

P

Paarl 69
Padova 30, 45, 46, 48, 61, 334
Pahlow, Ken 175
Paige, David 91
Parker, Robert 47, 274, 276
Parker Wong, Deborah ii, 273, 274, 276, 281
Parra, Sam 253, 263
Parra Wine Co. 253
Parr, Rajat 165
Passport to Pinot 89, 151, 152, 153, 154, 155, 157, 235, 291, 334, 345
Patricia Green Cellars 237
Paul, John 166, 167
Payne-Brown, Kate 178, 180
pedicel 213
Pedroni, Laura 224
peduncle 213
Pere Anselme 331
Perea, Omar 261, 263
Perkins Harter Wines 186, 190, 191
Perkins, Shelby 186, 190, 191, 194
Perry, Vivian 90, 94, 119
Peterson-Nedry, Harry 19, 89, 90, 97, 188, 280
Peterson-Nedry, Wynne 89, 207, 210, 280
petite sirah 34
petit mansang 308
petit verdot 69, 304
Petrus 295, 296
Peynaud, Émile 291, 322
Pfaffmann, G. 24
Pfarrkirche 22
Phelps Creek Vineyards 174, 328
Picpoul de Pinet 142
Piedmont 55, 133, 301
Piesporter Goldtröpfchen 23
Pigott, Stuart 85, 86, 87, 88, 89, 95, 99
pinotage 32, 33, 34, 68, 334
pinot blanc 226, 228, 279, 281
pinot gris 123, 173, 176, 179, 184, 200, 223, 226, 228, 275, 276, 280, 311, 334
pinot noir iii, 19, 30, 32, 33, 34, 37, 39, 43, 47, 48, 50, 58, 62, 63, 65, 74, 77, 86, 89, 93, 95, 112, 120, 121, 122, 123, 139, 151, 152, 153, 155, 156, 157, 158, 159, 160, 161, 162, 163, 165, 168, 173, 176, 179, 184, 200, 202, 203, 207, 210, 211, 213, 214, 215, 216, 217, 218, 219, 220, 223, 226, 228, 230, 231, 232, 234, 236, 237, 244, 246, 247, 262, 265, 274, 275, 276, 277, 278, 279, 280, 281, 290, 296, 303, 306, 307, 308, 309, 310, 311, 312, 314, 315, 319, 327, 328, 329, 333, 334, 335, 341
Plurality Voting 18
Pogue, Kevin 37, 38
Pollan, Christina 162
Ponnelle, Pierre 11, 296, 325
Ponzi, Anna Maria 91, 92, 119, 120, 121, 122, 123, 152, 160, 161, 170, 174, 255, 261, 263
Ponzi, Dick 92, 120, 121, 122, 248, 255
Ponzi, Luisa 92, 120, 124, 161, 170, 174
Ponzi, Michel 120
Ponzi, Nancy 120, 121, 122, 247, 248
Ponzi Vineyards 91, 92, 120, 122, 123, 174, 261, 263
Portugal 9, 51, 55, 125, 299
Potel, Nicolas 218
Pouilly-Fuissé 111
Precept Wine 202, 203
Premeaux-Prissey 291
Prešern, Domen 50
Prial, Frank 83, 122
Price-Quality Inversion 42
Princeton University 29, 30, 38, 50

Printed in Dunstable, United Kingdom

71020419R00218